大连自然博物馆
DALIAN NATURAL HISTORY MUSEUM

孟庆金　主编

文物出版社

北京·2006

主　　编：孟庆金
副 主 编：姜学品　刘兴和　栾淑君
撰　　稿：刘金远　孙　峰　李　梅　张淑梅
　　　　　孟庆金　赵永波　胡玉晶　高春玲
　　　　　黄文娟　程晓冬
图　　片：刘勤学　王暑杭
封面设计：隗　伟
图版摄影：郑　华
责任印制：张道奇
责任编辑：段书安

图书在版编目（CIP）数据

大连自然博物馆／孟庆金主编．—北京：文物出版社，
2006.5
ISBN 7-5010-1757-3
Ⅰ．大... Ⅱ．孟... Ⅲ．自然历史博物馆－简介－
大连市　Ⅳ．N282.313
中国版本图书馆CIP数据核字（2005）第053377号

大连自然博物馆

孟庆金　主编

文物出版社出版发行
（北京五四大街29号 邮政编码 100009）
http://www.wenwu.com
E-mail:web@wenwu.com
北京圣彩虹制版印刷技术有限公司制版印刷
新华书店经销
开本：889×1194 1/16 印张：14.5
2006年5月第一版　2006年5月第一次印刷
ISBN 7-5010-1757-3/N.3
定价：220元

目　录

序　Preface		单霁翔
前言　Foreword		孟庆金
大连自然博物馆概况　Dalian Natural History Museum Introduction		10
建筑　Architecture		10
沿革　Evolution		12
展示　Exhibition		14
教育　Education		16
藏品　Collection		18
科研　Scientific Research		20
交流　Exchange		21
地球：宇宙的骄子　Earth: The Pride of the Universe		22
宇宙遨游　Travel in the universe		22
陨石　Aerolites		23
地球漫步　Walk on the Earth		24
地球年龄的推断　An Inference of the Age of the Earth		24
古老的岩石　Old Rocks		24
生命的见证　Testimony for the Life		24
地下宝藏　Underground Treasure		25
金属矿产　Metal Minerals		25
特种非金属矿产　Special Nonmetal Minerals		25
非金属矿产　Nonmetal Minerals		26
药用矿产　Officinal Minerals		26
工艺美术原料矿产　Industrial Arts Raw Material Minerals		28
观赏石　Decoration Stones		29
大地沧桑　Great Changes of the Ground		30
火山活动　Volcano Activities		30
构造运动　Conformation Movements		31
鬼斧神工　The Wonderfulness of the Nature		31
人地和谐　The Harmony between Human and the Earth		32
化石：遥远的生命　Fossils: The Ancient Life		34
生命的出现与演化　The appearance and evolution of life		34
哺乳类　Mammals		37

鸟类　Aves	39
爬行类　reptiles	43
恐龙　Dinosaurs	43
翼龙　Pterosaurus	56
龟　Turtles	57
离龙　Choristoderes	58
两栖类　Amphibians	60
鱼类　Fishes	61
无脊椎动物　Invertebrates	63
虾类　Shrimps	63
蜘蛛类　Araneids	63
昆虫类　Insects	64
植物　Plants	65
模式标本　Type specimens	72

海洋：生命的摇篮　Ocean: The Cradle of Life　　76

海洋哺乳动物　Marine Mammals	77
海洋巨兽——鲸　Huge Animals in the Ocean-Whale	78
灵巧活泼的鳍脚类　Agile and Active Pinnipeds	99
美人鱼儒艮　Mermaid Dugong	103
海洋鱼类　Ocean Fishes	104
海洋软骨鱼类　Ocean Elasmobranch	105
海洋硬骨鱼类　Ocean Teleostean	117
海洋无脊椎动物　Ocean Invertebrates	134
多孔动物——海绵　Poriferan-Sponges	135
美丽的珊瑚　Beautiful Corals	136
多姿多彩的海洋贝类　Rich and Colorful Seashells	137
身穿"盔甲"的动物——甲壳动物　Animals with "Corselet"-Shellfish	145
海龟　Marine Turtles	146
藻类　Algae	147
美丽一族——红藻　Beautiful Group-Red Algae	147
经济植物之家——褐藻　Economic Plant Family-Brown Algae	149
潮间带的"绿金"——绿藻　Tidal Zone "Green Gold"-Green Algae	151

生物：多彩的世界　Biology: A Colorful World　　152

| 哺乳动物　Mammals | 152 |
| 　　哺乳动物的繁殖　Reproducing of Mammals | 155 |

哺乳动物的自卫	Self-defense of Mammals	157
哺乳动物的尾巴	Tails of Mammals	158
哺乳动物的牙齿	Teeth of Mammals	160
哺乳动物的飞行	The Flying of Mammals	161
高度进化的灵长类动物	Advanced Primates	161
食肉类哺乳动物	Flesh-eating Mammals	163
有蹄类哺乳动物	Ungulate Mammals	173

鸟类 Birds — 178
- 鸟类的主要特征　Main Features of Birds　179
- 形形色色的鸟类　Various Birds　182
- 神奇的湿地　Amazing Wetland　193
- 鸟类的迁徙　The Migration of Birds　196

两栖动物 Amphibians — 197
- 有尾两栖类　Amphibians with Tails　197
- 无尾两栖类　Amphibians without Tails　199
- 无足两栖类　Amphibians without Feet　199

爬行动物 Reptiles — 200
- 蛇　Snakes　200
- 蜥蜴　Lizards　201
- 鳄　Crocodiles　201
- 龟　Tortoises　202

珍稀淡水鱼类 Rare Freshwater Fishes — 203
- 鳇　Huso dauricus　203
- 肺鱼　Lungfish　203
- 其它淡水鱼类　Other Freshwater Fishes　204

昆虫 Insects — 205
- 蝶与蛾　Butterflies and Moths　206
- 甲虫　Beetles　212

植物 Plants — 217
- 蕨　Ferns　218
- 裸子植物　Gymnosperm　219
- 被子植物　Angiosperm　220

展览信息 — 228

序

自1905年实业家张謇先生创办南通博物苑开始，中国博物馆的发展已经走过一百年的路程。伴随着历史的沧桑巨变，博物馆事业发展迅速。全国现有博物馆2300多座，馆藏标本和文物2000多万件，每年举办陈列展览近10000个，接待观众1.5亿人次。这些辉煌成就，令世界瞩目。

自然史博物馆作为收藏、保护、研究和展示人类环境物证的文化教育机构，是一个国家宣传其文明成就和发展水平的重要窗口；是一个地区经济、社会进步的形象标志；是普及科学文化知识，提高公众科学文化素养和思想道德水平的教育基地。因此，自然史博物馆在区域社会中的影响越来越大，而随着人们文化需求的不断增长，博物馆为社会和社会发展服务的作用也越来越重要。

大连自然博物馆前身始建于1907年，是中国建馆最早的综合自然史博物馆。其历史悠久，馆藏丰富，新馆1998年对外开放，陈列荣获国家文物局"全国十大陈列精品奖"及"最佳新材料、新技术运用奖"。

本书从20多万件馆藏品中精选出1000多件标本，并配以30多幅陈列照片，以科普知识为线索，从地球、化石、海洋和生物四个方面，系统地介绍了大连自然博物馆的藏品和展示，思想新颖，图文并茂，是博物馆界不可或缺的又一部好书。

我衷心地祝贺这一新书的出版，并预祝大连自然博物馆为全国的博物馆事业做出新的、更大的贡献。

国家文物局局长
2006年4月8日

Preface

Chinese museum has experienced 100 years' development history since the industrialist Mr. Zhang Jian began to found Nantong Museum in 1905. With many great vicissitudes of history, the museum cause develops very fast. At present the whole country has more than 2000 museums with over 12 million specimens and cultural relics, and each year more than 8000 exhibitions are held, receiving 0.2 billion audiences/times. These splendor achievements obtained attracted the sight of the world.

As the cultural and educational institution that collects, conservations, researches and displays the material evidence of human being environment, natural history museum is an important window for a country to propaganda its civilization achievement and development level, an educational base to improve people's scientific and cultural accomplishment and thought and moral level. Therefore the influence of natural history museum in area society is getting bigger and bigger, and with the constantly increasing people's cultural needs, the role for museum to serve the society and the development of society becomes more and more important.

The former of Dalian Natural History Museum was founded in 1907, which is the earliest comprehensive natural history museum in China. It has a long history and rich collections. The new museum was opened in 1998, displaying "National Top 10 Display Exquisite Articles Prize" and "Best New Material, New Technology Application Prize" awarded by State Cultural Relics Bureau.

This book selected more than 1000 specimens from over 200,000 collections of the museum and illustrated with more than 30 display photos. Taking scientific popularization knowledge as a clue, it systematically introduces the collections and displays of Dalian Natural History Museum from four aspects of earth, fossil, ocean and biology, the thought is novel and both pictures and words are very attractive, really a good book that could not be ignored in museum industry.

I sincerely congratulate the publishing of this new book, and pre-congratulate that Dalian Natural History Museum would make new and greater contribution to the national museum cause.

Director general of State Cultural Relics Bureau

Shan Jixiang

April 8, 2006

前　言

世界上的发达国家，都拥有多个历史悠久的自然史博物馆。它不仅记录了一个国家，包括人类在内的生物界及其所生存的环境、演化与发展的真实情况，而且也是这个国家普及科学文化知识，提高公众科学文化素养和思想道德水平的教育基地。客观地说，中国的自然史博物馆不论在规模、数量上，还是在展示水平、教育职能的发挥上，都无法与国外的博物馆相比。就是在国内，自然史博物馆与其他类型的博物馆相比发展也相对滞后。随着社会、经济、文化的发展，自然史博物馆一方面要用收藏的标本为民族、历史、城市做文化的诠释，另一方面要满足社会和公众对教育、休闲、学习等方面的需求，所以自然史博物馆在区域社会中的地位以及在公众中的形象越来越重要。

自然史博物馆与其他类型的博物馆一样，其藏品是博物馆进行研究、展示和教育活动的基础，也是衡量博物馆规模和地位的重要标志之一。自然史博物馆的藏品主要是生物标本，标本的来源在二十世纪八十年代以前，主要是深入产地的广泛采集和动物园等相关机构及社会人士的友好捐赠，而目前则主要是靠征集。标本征集是受很多因素制约的。除了野生动物保护外，还要有资金、藏品来源信息、政策保障、高水平的征集人员、即时的价值判断，甚至还要有社会各界朋友的帮助。

自然史博物馆利用藏品进行的学术研究成果，可以为生命科学研究提供具体的佐证，形成知识与论点，也可以在自然资源调查、环境保护和文化交流方面做出贡献。把研究成果利用现代技术手段，通过标本和相关资料进行多元的展示，使知识得到推广和传播，这便是自然史博物馆所肩负的使命。

正因为藏品对于一个博物馆来说如此重要，所以，编撰这本以藏品为主要内容的图鉴，系统全面和生动地介绍大连自然博物馆近百年的收藏，这也是全国博物馆界和历代大连自然博物馆工作人员所期盼的。

大连自然博物馆有近百年的历史，经过历代博物馆人的不懈努力，在收藏、研究、展示和教育方面都取得了可喜的成果。本书从20多万件馆藏标本中，精选出1000多件各个时期具有收藏和研究价值的、保存完好的岩矿、古生物化石、植物、昆虫、两栖爬行类、鸟类、哺乳动物等珍稀标本，从地球、恐龙、海洋生物、东北森林、湿地等11个展厅中拍摄30多幅陈列照片，以照片、图形、文字相结合的表现形式，详细介绍了各种生物的形态、产地、分布、生态等方面的内容。在内容上，重点突出了海洋生物和"热河生物群化石"两大藏品特色，典型标本还辅以复原图、分布图等加以介绍，力求做到构图新颖、文字精练、表达准确、艺术性和可读性强，使读者能够系统地了解大连自然博物馆的藏品和"人与自然和谐"的展示主题。

愿本书能够成为各界人士了解大连自然博物馆的一个窗口，并给有志于从事博物馆研究的专家、学者提供有益的帮助，同时，也奉献给为本馆发展而呕心沥血的同仁们。

大连自然博物馆馆长　孟庆金

2006年4月8日

Foreword

All the developed countries in the world have several natural history museums with a long history, which not only records the true conditions of a country, the biological society including human being and their living environment, evolvement and development, but also the educational base for the country to popularize scientific and cultural knowledge, improve the people's scientific and cultural accomplishment and thought and moral level. Objectively speaking, Chinese museums could not compare with the museums abroad no matter form scale, quantity, or display level and exerting educational function. Even within the country, the development of natural history museums is backward than the development of other types of museums. With the development of society, economy and culture, on one hand the natural history museum would collect specimens to make an notation for the nation, history and city, on the other hand it should also meet the requirements and needs of the society and the people for education, leisure, study and so on, so the position of natural history museum in area society and its image among the people is getting more and more important.

Like the other types of museums the collection of natural history museums is the base for the museum to carry out researches, displays and educational activities, also one of the important marks to measure the scale and position of the museum. Mainly the collections of a natural history museum are the life-form specimens. Before 1980s the source of specimens are mainly to go to the origin to collect and the friendly donation of zoos and other relevant organs and social personalities, but at present it mainly relies on collecting. The specimen collecting is limited by many factors. Besides protection of wild animals, it also needs fund, collection source information, policy security, collecting persons with high level, immediate value judgment, even needs the help of friends from all circles of the society.

The learning research achievements obtained by natural history museum through collections study could supply concrete evidences for life science research, forming knowledge and lemma, as well as make contributions to natural resource survey, environment protection and cultural exchange. With the modern technical measures and through specimens and the relevant materials make the multi-display of the research achievements, making knowledge popularized and transmitted, which is the mission that a natural history museum shoulders.

Just as the collections are so important for a museum, so we composed this illustrated handbook with collections as its main contents, systematically, wholly and vividly introducing the collections of Dalian Natural History Museum for nearly hundred years, which is also what expected by the nation's museum's industry and the staff of Dalian Natural History Museum.

The history of Dalian Natural History Museum is nearly 100 years, through the constant efforts of the staff of the museum for several generations, we have made outstanding achievements in collection, research, display and education. From over 200,000 specimens of the museum this book carefully selected more than 1000 rare specimens of rocks, ancient life-form fossils, plants, insects, amphibians, birds, mammals and so on of each stage, which have the value of collection and research and also protected perfectly, takes more than 30 photos displayed from 11 exhibition halls mainly including earth, dinosaur, ocean life-form, Northeast forest, wetland and so on, in the combination form of photo, graph and words concretely introduced the contents of form, origin, distribution, zoology and so on of various kinds of life-forms. Regarding contents it mainly introduced the features of the two great collections of ocean life-form and "Jehol Biota Fossil", the typical specimens are also introduced with recovered pictures and distributive pictures and so on, so as to possibly make the pictures novel, words exquisite, expressions correct, strong artistry and readable, which enables readers to systematically realize the collections of Dalian Natural History Museum and the display theme of "Harmony of People and Nature".

We wish this book could become a window for personalities of all circles to understand Dalian Natural History Museum, and give a beneficial help to the experts and scholars who is willing to conduct the research of museum, meanwhile we also dedicate this book to our colleagues who work very hard for the development of our museum.

Curator of Dalian Natural History Museum
Meng QingJin
April 8, 2006

大连自然博物馆概况
Dalian Natural History Museum Introduction

建 筑 (Architecture)

　　大连自然博物馆是联合国教科文组织登记注册的大型自然史博物馆，是中国著名的自然史博物馆之一，其前身始建于1907年。旧址建筑是1898年沙俄修建的，具有浓郁俄罗斯风格，是大连市初建时期的代表性建筑之一，1997年被国务院列为国家重点文物保护建筑。

　　新馆建筑为典型的现代欧式风格，建筑面积15000平方米，1998年建成。坐落于风景秀丽的黑石礁海滨公园内，三面环海，礁石环绕，与著名的星海公园相邻。蓝色的屋顶与碧海、蓝天、白云相互辉映，景色怡人。在这里可以聆听大海与礁石的"对话"，可以垂钓、戏水……人与自然和谐的真谛，在这里得到了最好的诠释。

Dalian Natural History Museum is a large natural history museum first built in 1907 and registered at the UNESCO, which is one of the most famous natural museums in China.

The building at the original site of Dalian Natural History Museum was built by Russia in 1898 in a Russian

style, it is one of the masterpieces in the early stage of Dalian city, and the building was listed as one of the national key culture relic protection buildings by the state council in 1997.

The new building is in modern European style with a construction area of 15,000 square meters, the construction was completed in 1998. It is located in the scenic Heishijiao Costal Park, the museum is facing the sea in three directions and is embraced by reefs, and it is just next to the famous Xinghai Park. The blue roof is reflecting with the blue sea, azure sky and white clouds, and the scenery is marvelous. You can listen to the "dialogue" between the sea and the reef, you can go fishing and play in the sea... The true meaning of the harmony between human and nature is best embodied here.

大连自然博物馆旧馆
Dalian Natural History Museum Old Hall

沿革 (Evolution)

1907年，日本"南满洲铁道株式会社"在中国创办"地质调查所"，其是大连自然博物馆的前身。

1923年，"地质调查所"大量收集东北地区的自然标本和资料后，增设了陈列室，主要展示岩矿和部分古生物标本，并注明标本的产地、藏量、开采价值和用途等。

1926年，由于展示的标本种类增多，陈列内容增加，收集标本的地域不断扩大，涵盖了东北及蒙古等地的多种资源，将陈列室改为"满蒙物质参考馆"。

1928年"满蒙物质参考馆"在原建筑的两翼和后部扩建，又进一步将历年从我国东北、蒙古、西伯利亚、欧美等地搜集的岩矿标本以及农业、畜牧业、林业、水产业的实物标本和图文资料汇集起来，于同年11月份成立了供科学研究和观赏的"满蒙资源馆"。

1932年在展示原有资源标本外，还增加了东北、蒙古等地的民俗陈列内容，并将馆名改为"满洲资源馆"。

1945年大连解放后，8月23日由中国长春铁路公司接管，易名为"东北地方志博物馆"，中长铁路科研所委托苏联地质专家叶果洛夫担任馆长，并对原有的陈列进行修整。

1950年11月，中长铁路局将该馆移交给大连市人民政府文教局管理，同时将馆名改为"东北资源馆"，充实调整了陈列内容，主要展览我国东北地区的自然资源和建国后的新成就，成为向广大人民群众进行爱国主义教育、社会主义教育和普及科学知识的文化阵地。

1959年，正式定名为"大连自然博物馆"，并请中国科学院院长郭沫若先生题写馆名。

1995年，大连市为实施科教兴市战略，保护自然文化遗产，决定移址建设新馆。

1998年10月，新馆建成并对外开放。

In 1907, Japan "Southern Manchuria Railway Corporation" established "Geological Research Institution", which was the initial mold of Dalian Natural History Museum.

In 1923, after collecting a large amount of natural specimens and documents in northeastern China, the "Geological Research Institution" added display room mainly for rock ore and part of the ancient biology, the original place, storage amount, exploitation value and usages of the specimens were listed on the label for the visiting and research of a limited number of celebrities.

In 1926, with the increase of the specimens for display and the content of demonstration, the area where the specimens collected expanding to include various resources in northeast China and Mongolia, the display room was changed into "Manchuria and Mongolia Material Display House".

In 1928, "Manchuria and Mongolia Material Display House" expanded its construction by 2,000 square meters at the two sides and back of the original building, thus enriched its collection and tookin the rock ore specimens from northeast China, Mongolia, Siberia, Europe and America, the authentic specimens and the graphic and written documents in the fields of agriculture, stock raising, forestry and aquaculture, and the "Manchuria and Mongolia Resources House" was established in November of the same year for scientific research and visiting.

In 1932, the display of folk custom in northeastern China and Mongolia was added to the original resource specimen, and the name of the house was changed into "Manchuria Resources House".

After the liberation of Dalian in 1945, the museum was taken over by China Changchun Railway Station Company on August 23rd, and the name was changed into "Northeastern China Chorography Museum". Entrusted by the Zhongchang Railway Scientific Research Institution, the geologist (Eropob) from Soviet Union worked as curator and made some alterations to the previous display.

In Nov. 1950, Zhongchang Railway Station handed over the museum to the Culture and Education Bureau of Dalian municipality, at the same time the name was changed into "Northeastern China Resources House", the display content was enriched and adjusted, it was mainly for the exhibition of the natural resources of northeastern China and the new achievements made after the establishment of our country, the house became the culture base for patriotism, socialism and scientific knowledge education.

In 1959, its name was formally changed into "Dalian Natural History Museum", and the subscription was written by Guo Moruo, the dean of Chinese Academy of Science.

In 1995, with the implementation of the policy of prospering the city by science and education, Dalian City determined to relocate the museum to protect the natural and cultural legacy.

In Oct. 1998, the construction of the new exhibition hall was completed and it was opened to the public.

展示 (Exhibition)

本馆基本陈列共设有地球厅、恐龙厅、巨鲸厅、海洋哺乳动物厅、硬骨鱼厅、软骨鱼厅、海洋无脊椎动物与海藻厅、物种多样性厅、湿地厅、东北森林动物厅、辽西化石厅等11个展厅，共展出标本5,000余件。展示主题为"人与自然"。陈列展览的主要特色如下：

- 采用"主题单元展示法"布展；
- 以展示生物多样性为主要任务；
- 展览紧扣时代脉搏，强调保护环境；
- 陈列体现了最新研究成果；
- 树立"以人为本"的思想；
- 展览形式与内容设计和谐统一；
- 突出地方特色，为社区服务。

新馆陈列2001年荣获"全国十大陈列精品奖"及"最佳新材料、新技术运用奖"。

The demonstration area of the new exhibition hall mainly includes 11 halls: the earth hall, the dinosaur hall, huge whale hall, sea mammal hall, teleostean hall, elasmobranch hall, sea invertebrate animal and alga hall, species diversity hall, swamp hall, northeastern China forest animal hall and west Liaoning fossil hall, etc., the total number of specimens is more than 5,000, the main features of the exhibition are:

- The exhibition is disposed by method of "exhibition of the unit theme";
- We take the exhibition of species diversity as our main task;
- The exhibition is correspondent to the request of our time and lays an emphasis on environmental protection;
- The display embodies the latest research achievements;

湿地展厅
The Hall of Wetlands

- We hold "human orientation" as our principle;
- The harmony between the exhibition form and content;
- We give prominence to local characteristics and serve for the community.

The display of the new exhibition hall has won "Top Ten National Classic Display Prize" and "The Best Prize for the Utilization of New Material and New Technologies".

全国十大陈列精品奖证书
The certificate of the Top Ten National Classic Display Prize

硬骨鱼展厅
The Hall of Teleost

教育 (Education)

教育服务设施齐全，除了主题展示、解说导览、出版物、巡回展览、特别展览和学校教育推广外，还设有可供学术交流和进行文化活动的多功能厅，可以举办研讨会、演讲、培训、咨询服务、影片播放等；拥有27万平方米的天然海域，可供学生观察、了解海洋生物的行为和习性；配置了语音自动讲解系统、多媒体触摸屏、拼图活动等参与性项目，充分启发了观众的观察与思考；收藏各种专业、科普图书及档案资料9万余册，还有休息厅和观海台，给观众创造了一个方便舒适的学习和欣赏环境。

We have complete educational facilities including the multi-function hall for academic exchange and culture activities, the 270,000 square meters maritime space for students to observe and understand sea animals, the automatic voice explanation system, shop, relaxing room and sea view platform, more than 90,000 scientific books and documents in various specialties, more than 20 participating programs such as multi-media touch panel and jigsaw puzzle. All these will enlighten the observation and thinking of the audience, thus create a convenient and comfortable learning and visiting environment.

博物馆拥有的海域，是学生们的户外课堂。
The maritime space possessed by the museum is a perfect outdoor classroom for students

多媒体系统 Multi-media system

语音自动导览系统
Speech automatic-guiding system

多功能厅 Multi-function Hall

科研 (Scientific Research)

大连自然博物馆现有科研技术人员20多名,为博物馆的展览、收藏、科研、教育事业提供了保障。科研成果显著,已在各级各类刊物上发表专业论文数百篇,尤其是2004年9月,古生物学科研成果在世界著名科学杂志《自然》上发表。出版专著25本,荣获国家、省、市各级奖项多项。建有研究室、实验室、剥制室等,并配备各种仪器设备。

Dalian Natural History Museum has a strong technical team engaging in the museum career, which provides a guarantee for the exhibition, collection, scientific research and education of the museum. We have outstanding research achievements, including several hundreds of academic papers and 25 professional publications, and we have won many prizes at national, provincial and municipal level. The most remarkable achievement is that our research in the field of paleontology was published on the world famous Nature magazine in Sep. 2004, which is a contribution to the whole world.

日本、台湾和国内专家来馆合作研究 Co-operation with researchers from Japan, Taiwan and domestic scientists

贾兰坡先生来馆指导研究 Mr. Jia Lanpo superrised the researchers of the museum

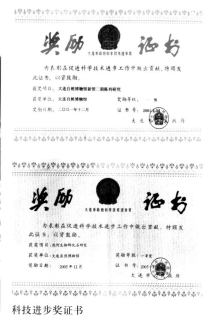

科技进步奖证书
Science and Technology progress prize

非洲展厅陈列设计研究
Assess the design of African animal hall

《自然》杂志 Nature magazine

与美国专家进行化石研究 Study focsils with American erperts

藏品 (Collection)

　　馆藏包括动物、植物、古生物、岩矿、人类学标本等各种珍稀标本近20万件，以及相关文物、照片、影像资料、野外记录、档案资料等。其中大型海洋生物标本和完整的"热河生物群"化石标本是馆藏的突出特色。包括重达66.7吨的黑露脊鲸、50多吨的抹香鲸等20多种海兽标本和中华龙鸟、尾羽龙、神州鸟等带羽毛的恐龙化石标本。库房采用新式密集架，标本使用藏品管理信息系统实现计算机化管理。

The collection includes abundant specimens such as animals, plants, ancient biology, rock ore and sea animals, the total number of the rare specimens is nearly 200,000, among which the large sea animal specimen and complete "Rehe Biome" are the characteristics of our museum. The collection also includes more than 20 specimens of sea animal such as Northern right whale with a weight of 66.7 tons, sperm whale with a weight more than 50 tons and the feathered dinosaur fossils specimen of Sinosauropteryx, Caudipteryx, Shenzhouraptor, etc. The management information system of the standard storage house is administered by computer, which has increased our work efficiency and quality.

藏品管理信息系统
Collection management information system

库房
Storage house

交流 (Exchange)

　　加强对外交流工作，提供馆藏标本与国内外学者进行学术研究；与世界著名的大学、博物馆等相关机构建立良好的合作关系；与美国、日本等国家的博物馆、展览机构等进行了展览和学术交流；进一步提升本馆在国内外的地位。

We give great attention to communication with appropriate overseas institutions. We have established good relationships with State Natural History Museum of France, Kitakyusyu Natural History Museum of Japan, and have conducted many exhibitions and academic exchanges with universities and museums in U.S.A and Japan.

与日本北九州自然博物馆交流
Communicate with the Kitakyushu Natural History Museum of Japan

在日本举办东北地区鸟兽展
Northeastern birds and animals exhibition in Japan

在加拿大举办古生物化石展
Fossils exhibition in Canada

在日本举办化石展
Fossils exhibition in Japan

宇宙遨游 (Travel in the universe)

目前观测到的宇宙是由大约1000亿个星系组成的,每个星系拥有上千亿个恒星。星系聚集成星系团,星系团又集结成更大的超星系团或星系云,而超星系团则远没有填满宇宙,宇宙还在不断地膨胀着。

The observable universe is composed by about 100 billion galaxies, every of which has stars up to 100 billion. Galaxies gather into galaxy groups, the groups will converge into more gigantic super-galaxy groups or nebula, however, the number of super-galaxy groups is too small to cram the universe, which is continuing to expand.

宇宙星云图

Nebula in the universe

陨石 (Aerolites)

陨石来自遥远而古老的太空，陨石上记录着50亿年来太阳系演变的证据。陨石主要来源于小行星和彗星，大体上可分为三大类：铁陨石、石陨石和铁石陨石。

Aerolites come from the remote and ancient space, the evolution of the solar system in the past 5 billion years is recorded in the aerolites. Aerolites mainly rooted in asteroid and comet; there are generally three types of aerolites: siderolite, stone aerolite and stone siderolite.

光明山石陨石
1998年陨落于大连市庄河境内

Guangmingshan Stone Aerolite Fell from the sky into Zhuanghe, Dalian city in 1998

庄河石陨石
1976年陨落于大连市庄河境内

Zhuanghe Stone Aerolite Fell from the sky into Zhuanghe, Dalian city in 1976

海南玻璃陨石
产自海南

Hainan Glass Aerolite (Hainan)

吉林石陨石
1976年陨落于吉林省吉林市境内

Jilin Stone Aerolite Fell from the sky into Jilin city, Jilin Province in 1976

乌珠穆沁铁陨石
产自内蒙古

Ujimqin Siderolite (Inner Mongolia)

地球展厅
the Earth Exhibition Room

地球漫步 (Walk on the Earth)

地球年龄的推断 (An Inference of the Age of the Earth)

推算地球年龄,主要依据岩层方法、化石方法和放射性元素的蜕变方法等。根据鉴定,地球上最古老的岩石,是格陵兰岛的阿米佐克片麻岩,约38亿年。月球岩石的年龄是46亿年,陨石的年龄都在45～47亿年之间,据此推断,地球的年龄应为46亿年。

The inference of the age of the earth takes terrane method, petrification method and radioelement spallation method. According to the appraisal, the oldest rock on the earth is the Umiivik gneiss in Greenland, its age is about 3.8 billion years. The rock on the moon has an age of 4.6 billion years, the ages of the aerolites on the moon are between 4.5 and 4.7 billion years, thus the age of the earth should be 4.6 billion years.

古老的岩石 (Old Rocks)

硅质页岩
产自河北,距今约18亿年
Siliceous
From Hebei About 1.8 billion years ago

砾岩
产自河北,距今约17亿年
Conglomerate shale
From Hebei About 1.7 billion years ago

石英砂岩
产自河北,距今约15亿年
Quartz sandstone
From Hebei About 1.5 billion years ago

生命的见证 (Testimony for the Life)

多轮辽南水母
产自大连地区,大约8亿前
Acalephe
From Dalian Area About 800 million years ago

三叶虫
产自大连地区,大约5亿年前
Trilobite
From Dalian Area About 500 million years ago

转角羚羊
产自大连地区,大约1万年前
Antelope
From Dalian Area About ten thousand years ago

地下宝藏 (Underground Treasure)

金属矿产 (Metal Minerals)

金属矿产是指能够供冶金工业提取各种金属的矿物。

It refers to minerals that can be used in metallurgy industry for metal.

斑铜矿
产自吉林

Bornite (From Jilin)

黄铜矿
产自大连

Chalcopyrite (From Dalian)

黑钨矿
产自湖南

Wolframite (From Hunan)

方铅矿
产自辽宁

Galena (From Liaoning)

钨锰铁矿
产自广东

Wolframite (From Guangdong)

磁铁矿
产自辽宁

Magnetite (From Liaoning)

自然金与试金石
产自大连

Gold and touch stone (From Dalian)

赤铁矿
产自辽宁

Hematite (From Liaoning)

特种非金属矿产 (Special Nonmetal Minerals)

特种非金属矿产是指主要利用其特殊物理性质，如电学、光学、研磨、绝热、绝缘、隔音等方面特性的矿产。

It refers to the mine products whose physical properties such as electricity, optics, rubbing, heat proof, insularity and sound proof can be utilized.

白云母
产自大连

Muscovite (From Dalian)

温石棉
产自大连

Chrysetile (From Dalian)

冰洲石
产自浙江

Iceland spar (From Zhejiang)

水晶
产自辽宁

Rock Crystal (From Liaoning)

非金属矿产 (Nonmetal Minerals)

非金属矿产是指能供工业提取有用非金属元素、化合物或能直接利用的岩石和矿物集合体。

It refers to the aggregation of rock and minerals that can provide nonmetal elements or compounds to industries or can be utilized directly by industries.

菱镁矿
产自辽宁

Magnseite
From Liaoning

萤石
产自浙江

Fluorite
From Zhejiang

油页岩
产自辽宁

Oil Shale
From Liaoning

抚顺煤
产自辽宁

Fushun Coal
From Liaoning

药用矿产 (Officinal Minerals)

药用矿产是指可作为药用的天然矿物、岩石和生物化石等。
It refers to the natural mineral, rock or biological fossil can be used as ingredients for medicine.

朱砂
矿物名辰砂，产自贵州
古代练丹的主要原料，
具有镇惊、安神等功效。

Vermilion
Mineral name Cinnabar, the produce area is Guizhou, it is the main material for ancient medicine-making with the functions of pacifying and soothing the nerves.

铜绿
矿物名孔雀石，产自俄罗斯
具有解毒、收敛、杀虫等功效。

Patina
Mineral name Malachite, the produce area is Russia, its main functions are detoxification, constringency and killing insects.

雌黄
矿物名雌黄，产自湖南
具有解毒、杀虫等功效。

Orpiment
Mineral name Orpiment, the produce area is Hunan province, it has the functions of detoxification and killing insects, etc.

雄黄
矿物名雄黄,产自湖南
具有解毒、杀虫等功效

Realgar
Mineral name Realgar, the produce area is Hunan province, it has the functions of detoxification and killing insects, etc.

滑石
矿物名滑石,产自辽宁
具有清热解暑、利尿渗湿等功效

Talc
Mineral name Talc, the produce area is Liaoning province, it has the functions of heating clearing, diuresis and moisture elimination.

炉甘石
矿物名菱锌矿,产自广西
具有明目去翳、生血生肌等功效

Calamine
Mineral name Smithsonite, the production area is Guangxi Province, it has the functions of brightening the eyes and blood generation.

地球展厅
the Earth Exhibition Room

工艺美术原料矿产 (Industrial Arts Raw Material Minerals)

是指符合工艺美术要求可开发利用的矿物单晶体、集合体。
It refers to the single crystalloid and aggregation that meet the exploitation requirement of industrial arts.

黄玉
产自俄罗斯

Topaz (From Russia)

孔雀石戒面
产自湖北

Malachite (From Hubei)

猫眼石
产自俄罗斯

Cat's Eye (From Russia)

翡翠原料及戒面
产自缅甸

Jade material and Jadeite (From Burma)

电气石
产自俄罗斯

Tourmaline (From Russia)

煤精工艺品和平鸽
产自辽宁

Craftwork Peace Pigeon made of jet (From Liaoning)

大连市瓦房店金刚石矿
Dalian Wafangdian Diamond Field

金伯利岩中的金刚石
产自大连

Diamond in Kimberley
From Dalian

观赏石 (Decoration Stones)

观赏石是自然界中外形奇特，色泽艳丽，纹饰美观，不经加工即具有观赏、玩味、收藏价值的矿物、岩石和生物化石的总称。

It is the general name of rock and biological fossils such as minerals with particular shape and beautiful color and patterns, and are worthy of appreciation, relish and collection without any processing.

紫水晶
产自大连

Amethyst (From Dalian)

辉锑矿
产自湖南

Antimonite (From Hunan)

萤石晶簇
产自湖南

Fluorite cluster (From Hunan)

长石晶簇
产自朝鲜

Orthoclase cluster (From Korea)

黄铁矿
产自广东

Pyrite (From Guangdong)

雌黄
产自湖南

Orpiment (From Hunan)

水晶晶簇
产自海南

Crystal cluster (From Hainan)

石灰华
产自山西

Calc-tufa (From Shanxi)

大连自然博物馆

大地沧桑 (Great Changes of the Ground)

　　地壳自形成以来，就受到强大的内、外力作用，使其表面形态、组成物质和内部结构都在不停地运动、变化和发展着。岩石的变形、海陆的变迁及千姿百态地表形态，都是地壳变动的结果。

Since the formation of the crust, it suffers from strong internal and external forces, which makes its surface configuration, substances and internal structure move, change and develop continuously. The distortion of rocks, changes of sea and lands and various earth surface configurations are all the results of crust change.

地貌（地球展厅一角）
A Part of the Earth Exhibition Hall

火山活动 (Volcano Activities)

　　火山爆发喷出的熔岩，有的在地表堆起高山，有的在海洋中造成新岛。
Some of the lava from volcano eruption piles up as high mountains in the ground, and others form new islands in the sea.

火山弹（火山喷发使岩浆迅速冷却而形成，产自黑龙江）
Volcanic Bomb
(From Heilongjiang, It is formed by the lava which is cooled quickly after the eruption)

火山喷发形成的新岛
New Islands Formed after the Eruption

构造运动 (Conformation Movements)

　　水平运动使岩层发生水平位移和弯曲变形，常常造成巨大的褶皱山系；升降运动使岩层表现为隆起或凹陷，引起地势的高低起伏和海陆变迁。

Horizontal movement makes the rock layer move in horizontal direction and distortion, thus often creates huge drape mountain systems; the lifting movement makes the rock layer raise or recess, and creates the ups and downs of the physiognomy and the changes of the land and sea.

构造运动使岩石发生褶曲，产自辽宁

Drapes on the rock formed in conformation movement
(From Liaoning)

断层的挤压使岩石产生光滑的镜面，产自河北

Slippery surface on the rock because of the extrusion of the fault
(From Hebei Province)

鬼斧神工 (The Wonderfulness of the Nature)

　　内力作用为地表形态建造了"粗毛坯"，大自然的鬼斧神工——风化、剥蚀、搬运、沉积、成岩等外力作用，进而把地表形态雕塑得多姿多彩，美丽如画。

The internal force created the "roughcast" of the ground surface configuration, the unbelievable skills of nature-efflorescence, denudation, replacement, aggradations and rock formation have sculpted the ground surface into such a changing and beautiful landscape.

波痕，水的流动在泥沙中留下的痕迹，产自大连

Ripple Mark
(The marks left by the flowing water in the sand, From Dalian)

钟乳石，由于溶洞内富含碳酸钙的成分而使钟乳石象结石一样生长，产自河北

Stalactite
(The calcium carbonate rich in the cave makes the stalactite grow like calculi, from Hebei)

人地和谐 (The Harmony between Human and the Earth)

热带雨林被砍伐、现代生物物种的迅速灭绝、空气和水到处发生污染、全球变暖、以及臭氧层出现空洞等，使地球上的生态系统受到了前所未有的破坏，也给人类的生存带来了威胁。我们只有一个地球，她是我们赖以生存的家园。

The rain forest chopped down, the quick extinction of species, the pollution of the air and the water, global warming and the hole in the ozone layer are destroying the eco-system on the earth in an unprecedented scale, and have brought threats to the existence of human being. We have only one earth and we are depending on her for living.

干旱
Drought

遭受污染的土地无法耕种
The polluted soil is not suitable for culture

海滩上的垃圾
Garbage on the beach

污水灌溉使土壤板结
Hardened soil due to polluted water irrigationto

风沙入侵林区
Sand evading the forest

正向天空排放着废烟的工厂
Factory disposing smoke to the sky

铁路沿线的垃圾
Garbage along the railway

酸雨造成森林枯萎
Wilted forest after acid rain

有害的垃圾每天都在生产
Everyday the harmful garbage is in production

水质污染无法使用
The water is polluted and cannot be utilized

森林砍伐
Forest felling

过多的人口使地球承受巨大的压力
The overburden earth due to population explosion

龙卷风
Tornado

干旱使动物和植物大量死亡
The death of large amount of animals and plants due to drought

受污染水中的生物大面积死亡
The large-scale death of the animals in the polluted water

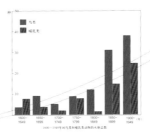

物种在逐年消失
The gradual disappearance of species

化石——遥远的生命
Fossils: The Ancient Life

生命的出现与演化 (The appearance and evolution of life)

据可靠文字记载，地球上最早的生命出现在距今35亿年前，是一种名为古球菌的炭质球粒，在生命出现的初期，生物的演化极其缓慢。直到距今6亿年前，所发现的都是细菌和藻类、叠层石等原始生命化石。

According to the reliable record, the first creature came to appearance 3500 million years ago, which was a kind of charcoal grains named as Archarosphaeroides. The evolution speed was very slow during the primary life period. There were all bacterium, algae and other primary lives 600 million years ago.

叠层石 Stromatolite

在距今大约6亿年左右，生物演化出现了质的飞跃，先是带硬壳的动物趋向繁荣，即而脊索动物出现，并登陆成功，植物中则先是出现了陆生的低矮灌木丛，接着高大的乔木植物石松类和有节类代替了矮小的灌木，呈现出水陆生物并行发展的面貌。

There was an essential change about 600 million years ago. Firstly, the chordates with hard shells boomed. They appeared in large amounts and successfully came to the continents. Secondly, the lower shrubs dominated the lands and then the higher arbors took their places. The earth showed a lively appearance both in water and on lands.

三叶虫 Trilobite

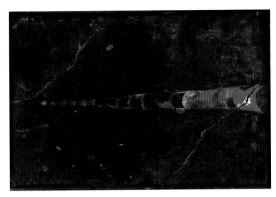
直角石 *Orthoceras*

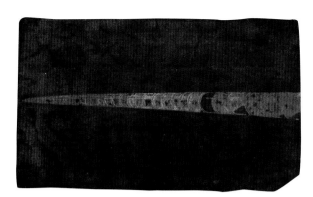
中国直角石 *Orthoceras sinensis*

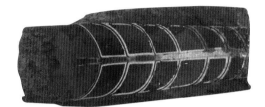

珠角石 *Actinoceras*

海百合 Crinoidea

海胆 Echinozoa

鸭头贝 *Stringocephalus*

生活时代：中泥盆世（距今 3.87～3.74 亿年）

Geological Era: Middle Devonian (387~374 million years from today)

进入中生代，生物界的面貌发生了翻天覆地的变化，首先是菊石类、穿孔类、腹足类等无脊椎动物的出现，特别是菊石类的发展，经过了从简单到复杂的过程，最后随着中生代的结束也绝灭了。

The life world changed a lot in Mesozoic era. The ammonites、the gasteropods and other invertebrates came to appearance, especially the ammonites lived through a period from simple to complex, and became extinct at the end of Mesozoic era.

菊石　Ammonoidea

中生代也是脊椎动物繁荣的时代，尤以爬行动物处于极盛，鸟类和哺乳类开始出现并趋向发展，鱼类也有了相当的数量。

陆地植物在彻底摆脱了对水的过分依赖后，出现了耐旱的植物——裸子植物和被子植物。特别是进入新生代之后被子植物迅速崛起并快速发展，把地球打扮成五彩缤纷的世界。

出现于中生代末期的哺乳动物，进入新生代后迅速的发展，代替爬行类成为地球的新主人。

Mesozoic era was a prosperous period for vertebrates. The reptiles reached their peak time, birds and mammals appeared and developed rapidly, fish also played an important part.
The drought-resistance plants-the gymnosperms and the angiosperms came to appearance after the terrestrial plants got independent of the water. Especially the angiosperms developed rapidly in Cenozoic era, they decorated a colorful earth. The mammals appeared at the end of the Mesozoic and became the domination after taking the reptiles' place.

哺乳类 (Mammals)

哺乳动物是动物界中最高等的一类。其最大的特征是胎生、哺乳。
Mammalia is the highest among animals. The most distinguishing characters are viviparous and suckling.

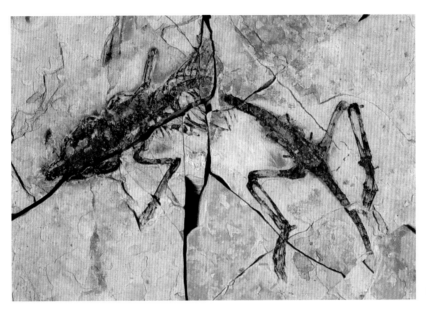

多瘤齿兽 Multituberculata

披毛犀头骨 Skull of *coelodonta antiguitatis*

原始牛角 Horn of *bos primigenius*　　　　东北野牛头 Head of *bison exiquus*

海象牙 Teeth of *odobennus*

猛犸象臼齿 Moldar of *mammuthus*

猛犸象动物群复原图
Recovered Picture of *mammoth*

鸟类 (Aves)

鸟类是一类身披羽毛的恒温飞行动物。与飞行的功能相适应，它的骨骼系统中出现了与之相应的变化。对于鸟类起源的研究，有专家认为鸟类就是会飞的恐龙。

Aves is a kind of warm blood winged animal, covered with feathers. They derived lots of changes adapting the flight. Some researchers believe that the birds are derived from dinosaurs.

神州鸟 (*Shenzhouraptor*)

神州鸟是迄今为止在中国境内发现的最原始的鸟类。该标本全身羽毛印痕保存完好、清晰。化石全身长53公分，没有牙齿，尾巴较长，具有了一定的飞行能力。

Shenzhouraptor is the oldest raptor found in China until now. The feather moulage of the specimen is complete and clear. The total length of the fossil is 53 cm, it has no tooth. Its tail is relatively long which enables the raptor to fly.

神州鸟 *Shenzhouraptor* sp.

圣贤孔子鸟 (*Confuciusornis sanctus*)

孔子鸟是已发现的最早由角质喙代替牙齿的原始鸟类。其最大的特征是肱骨近端非常扩展，横向特别宽，且近端中区有一椭圆形的比较大的气囊孔。

孔子鸟是一种飞行能力较强的鸟类，是继始祖鸟134年之后世界上发现的第2种原始鸟类化石。它的发现对于研究与探讨鸟类的起源与演化具有深远的意义。

Confuciusornis sanctus is the oldest bird whose teeth were replaced by cutin beaks that we can find. The most obvious feature is that the proximal end of the humerus is expanded in a large scale and the traverse side is very broad, there is a distinctive fenestra on the proximal end of the humerus.

Confuciusornis sanctus has a strong flying capacity, it is the second original bird fossil after the discovery of *Archaeopteryx* fossil 134 years later. This discovery has a significant meaning to the research and exploration of the bird origin and evolution.

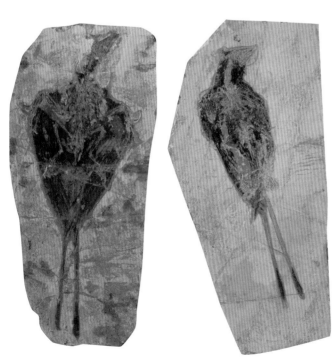

圣贤孔子鸟 *Confuciusornis sanctus*

圣贤孔子鸟复原图
Recovered Picture of *Confuciusornis sanctus*

杜氏孔子鸟 (*Confuciusornis dui*)

杜氏孔子鸟具有清晰的角质喙印痕和明显的双弓型头骨。其特征是下颌骨前部比较细；胸骨前缘中央有一呈"V"字型的凹刻构造，胸骨两侧有一对短的后侧突；尾综骨末端膨大。

The skull of *Confuciusornis dui* is of tycical diapsid type, and horny beak is clearly preserved. Its features are that the mandible is slender anteriorly.The sternum has a "V" shape concave structure and a pair of short latero-posterior processes , The pygostyle is expanded distally.

杜氏孔子鸟复原图
Recovered Picture of *Confuciusornis dui*

杜氏孔子鸟 *Confuciusornis dui*

反鸟类 (Enantiorhithes)

反鸟是因为它的某些身体构造与现生鸟类正好相反而得名。始反鸟是当时已知的最早的反鸟类化石。

It is named so because certain part of its body is just the opposite to present birds. *Eoenantiornis* was the oldest birds when it was discovered then.

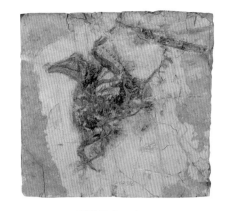

长翼鸟 *Longipteryx*

始反鸟复原图
Recovered Picture of *Eoenantiornis*

始反鸟 *Eoenantiornis buhloris*

今鸟类 (Ornithurae)

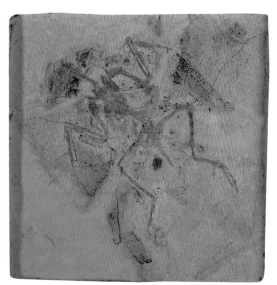

义县鸟 *Yixianornis*

爬行类 (Reptiles)

恐龙 (Dinosaurs)

恐龙是古生物学中最引人入胜的一类早已绝灭的爬行动物,它的发生、发展和绝灭也是生命进化史上最为动人的一章。有关它的奥秘一直是科学家热衷于探讨的课题。

Dinosauria is one of the most interesting extinct reptiles, its origin, development and extinction is a very charming chapter in the life evolution. All things about dinosaur are always the focus chased by the scientists.

恐龙展厅 The Hall of Dinosaur

合川马门溪龙 (*Manmenchisaurus hochuansis*)

合川马门溪龙是晚侏罗世时期植食性的大型恐龙。其最大的特点是头小，脖子特别长，有19节颈椎。

Manmenchisaurus hochuansis is a large plant-eating dinosaur in late Jurassic. Its particular features are that it has a small head and a long neck with 19 neck vertebra.

合川马门溪龙
Manmenchisaurus hochuansis

马门溪龙复原图
Recovered Picture of *Manmenchisaurus hochuansis*

棘鼻青岛龙 (*Tsintaosaurus spinorhinus*)

棘鼻青岛龙是新中国成立后首次发现的完整恐龙化石，属于植食性鸭嘴龙类。因发现于我国青岛市，又有棘鼻状的顶饰而得名。

Tsintaosaurus spinorhinus is the first completed dinosaur fossil found after the establishment of the People's Republic of China, it is plant-eating Hadrosauridae. It was named so because it was found in Qingdao, and there is spinoblast shape adornment on its head.

棘鼻青岛龙复原图
Recovered Picture of *Tsintaosaurus spinorhinus*

棘鼻青岛龙 *Tsintaosaurus spinorhinus*

巨型永川龙 (*Yangchuanosaurus magnus*)

巨型永川龙是一种大型肉食型恐龙。体长9.5米，站起来有5米多高。它的头骨高大粗壮，眼孔很大，说明它视力很好。较长而粗壮的后肢说明它主要是靠后肢走路。

Yangchuanosaurus magnus is a large flesh-eating dinosaur with a length of 9.5 meters and a height of 5 meters when standing. Its skull is high and strong with large eyelet in it, which can explain that it has a very good vision. The long and strong back limbs testify that they mainly walk on back limbs.

巨型永川龙复原图
Recovered Picture of *Yangchuanosaurus magnus*

巨型永川龙 *Yangchuanosaurus magnus*

沱江龙（*Tuojiangosaurus*）

多棘沱江龙是剑龙的一种。其特征是尾部有四根棘刺，背上有两行骨质的三角板，用以防御敌害和保护身体。

Tuojiangosaurus multispinus is a kind of Stegosauridae. It has four thorns on its tail, two bone set square on its back for defense and protection of the body.

多棘沱江龙 *Tuojiangosaurus multispinus*

恐龙蛋（Dinosaur Eggs）

恐龙蛋的埋藏形式与古环境学有密切的关系，为研究探索恐龙的行为学、生理学提供了资料。恐龙蛋是依据蛋壳的形态构造来分类的。

The burying way of dinosaur eggs has a close relationship with ancient environmental protection and has provided documents for the research of dinosaur behavior and physiology. The dinosaur eggs are classified according to the formation of the shells.

恐龙蛋 Egg of dinosaur

甲龙 (Ankylosauridae)

甲龙是恐龙防御系统演化最完善的一类，除了头部被骨甲包围外，身体从脖子到尾巴，都有甲板保护着，并且还长有尾锤。

Ankylosauridae has the most advanced dinosaur defense system, its head is enclosed by bone shell and there are shells on its body from head to tail, in addition, it also has hammer-shaped tail tip.

甲龙复原图 Recovered Picture of Ankylosauridae

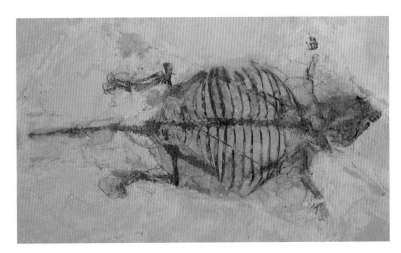

甲龙 Ankylosauridae

辽西化石展厅
Western Liaoning Fossil Exhibition Hall

鹦鹉嘴龙 (*Psittacosaurus*)

　　鹦鹉嘴龙是东亚地区早白垩世特有的一类小型恐龙。头呈三角形,有向外突出的颧骨突,吻尖呈钩状如鹦鹉的嘴。

Psittacosaurus is a small dinosaur only lived in East Asia in early Cretaceous period. It has triangle head with protruding cheekbones and hooked proboscis tips like the mouth of parrot.

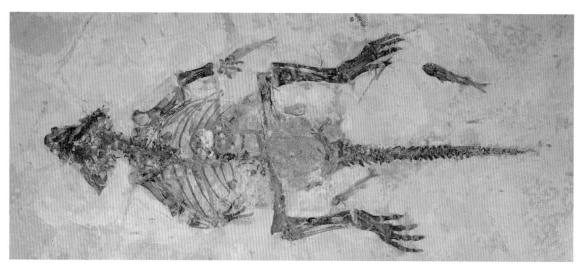

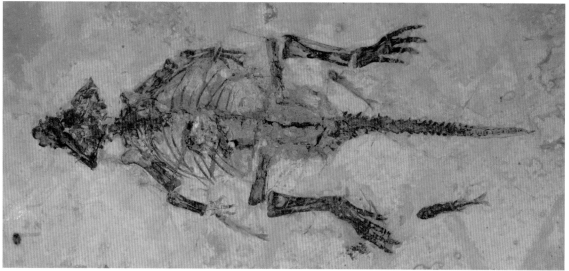

鹦鹉嘴龙 *Psittacosaurus*

这是一件世界上罕见的、迄今为止所发现的个体数量最多的一窝恐龙化石标本,由34条鹦鹉嘴龙幼体和1个成年鹦鹉嘴龙个体组成。幼体平均长度在23厘米左右,其头骨愈合疏松,骨缝明显,反映出是一群刚出生不久的幼仔。它的发现证明:恐龙具有育幼行为。

This is the rare dinosaur nest specimen fossil with the largest amount of dinosaurs till now, and it is composed of 34 minor *Psittacosaurus* and one adult *Psittacosaurus*. The average body length of the minors is 23cm and their skull connections are loose with obvious bone slot, we can infer that the minors were just born for a short time. This discovery proved that dinosaurs have minor breeding behaviors.

鹦鹉嘴龙复原图
Recover Picture of *Psittacosaurus*

中华龙鸟 (*Sinosauropteryx*)

中华龙鸟是一种小型兽脚类恐龙，其刀状的牙齿有锯齿形的边缘。腰带三射型，尾长。两足行走，从头到尾尖有一列表皮衍生物，尚不具备羽毛的特征，处于羽毛进化的初级阶段，只能称其为"原羽"或"前羽"。

Sinosauropteryx is a small theropoda dinosaur, its knife-shaped teeth have sawtooth edge. It has a triradial pelvic gridle with a long tail. It walks on two feet, there are epidermis ramification without any feather features covering its body from head to tail, thus they are still in the initial phase of feather evolution, the feather like can only be called as "Profeather" or "Primality feather ".

中华龙鸟复原图
Recovered Picture of *Sinosauropterys*

中华龙鸟 *Sinosauropterys* sp.

尾羽龙 (*Caudipteryx*)

尾羽龙是一种具有真正羽毛的恐龙。其牙齿退化,胃部保留着胃石,用以帮助消化。其不对称的羽毛尚不具备飞行的功能,是羽毛演化相对原始的阶段。

Caudipteryx is a dinosaur with real feather. Their teeth are degenerated and there are bezoars in their stomach to help digest. The asymmetrical feather does not enable the dinosaur to fly, and it is still in the prima period of feather evolution.

尾羽龙复原图
Recovered Picture of *Caudipteryx*

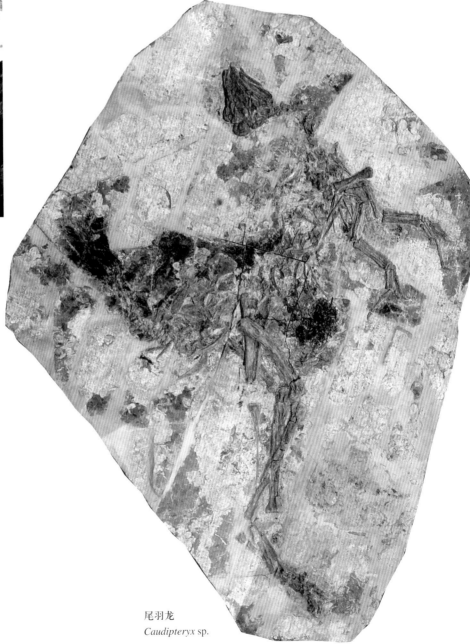

尾羽龙
Caudipteryx sp.

小盗龙 (*Microraptor*)

小盗龙是小型兽脚类恐龙,牙齿锋利,四肢长有羽毛。它最大的特点是在与身体长度相当的棒状尾巴上,布满了密密麻麻的筋腱。

Microraptor is a small theropoda dinosaur with sharp teeth, there is feather on their limbs. The particular feature of this dinosaur is that the tail which is almost in the same length as the body is covered with thick tendons.

小盗龙复原图
Recovered Picture of *Microraptor*

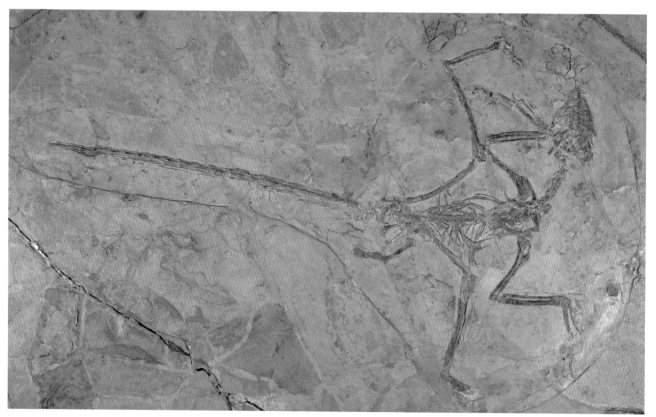

小盗龙 *Microraptor* sp.

中国鸟龙 (*Sinornithosaurus*)

中国鸟龙在分类上属于兽脚亚目的驰龙科，它的前肢已经不再像多数兽脚类恐龙一样向前腹向伸展，而是向鸟类的翅膀一样向上向侧面伸展。这一转化为鸟类飞行的起源在骨骼结构上奠定了基础。

Sinornithosaurus is belong to Dromaeosauridae, Theropoda. Its forelimbs extend toward two sides like the wings of bird rather than toward ahead-abdomen like most theropoda. This change establishes foundation for flying origin of bird on structure of bones.

中国鸟龙 *Sinornithosaurus* sp.

翼龙 (Pterosaurus)

翼龙类属于飞行爬行动物，是最早飞上天空的脊椎动物，是恐龙的远亲近邻，也是中生代的空中霸主。

Pterosauria belongs to the winged reptiles. It was the earliest vertebrate which could fly in the air. It's the near neighbour of dinosaur, but they are Scotch cousin (distant relatives). It was also the aerial domination of Mesozoic era.

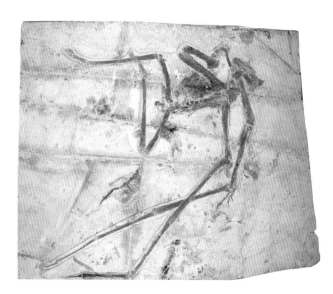

翼龙 Pterosaurus

龟 (Turtles)

龟类是现生动物中比较特化的一支，它的躯干包藏在骨质甲壳内，这使它迥异于其它的动物。同时它也是具有较长演化历史的一支。关于它的起源问题，至今仍然是个未解之谜。

Tortoise is a special group among current animals. Its body hides inside the ossified deck, this make it distinguish from other animals. It is one of the groups which have long evolution history. It is still a riddle about its origin.

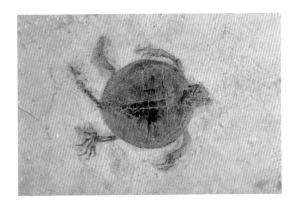

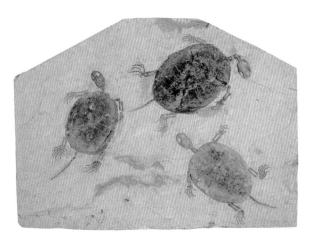

满洲龟 *Manchuensis*

离龙 (Choristoderes)

离龙类是一类单系的水生爬行动物。百余年来，人们对此类群了解甚少。
Choristodere is a single-derived aquatic reptile. They were poorly known in the last hundred years.

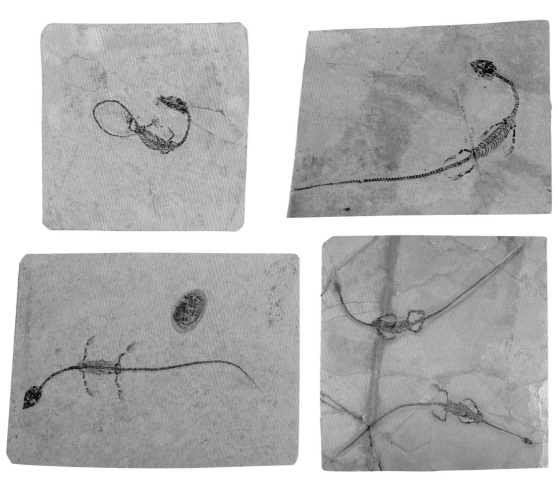

凌源潜龙 *Hyphalosaurus lingyuanensis*

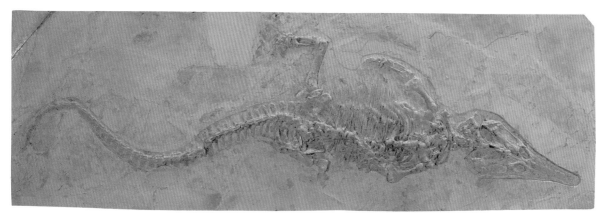

伊克昭龙 *Ikechosaurus*

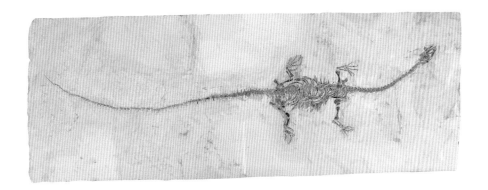
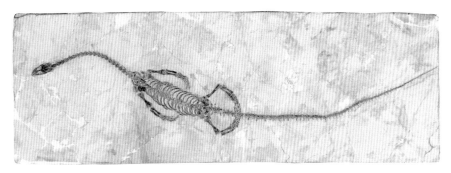

凌源潜龙 *Hyphalosaurus lingyuanensis*

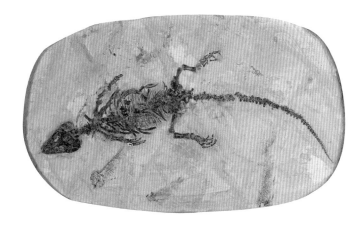

满洲鳄 *Monjurosuchus splendens*

两栖类 (Amphibians)

在脊椎动物的进化史中,从水生到陆生是一次重要的飞跃,两栖类就是代表这一演化阶段的过渡类群。与其他脊椎动物相比,两栖类的化石十分稀少。

During the vertebrate evolution, it was an important change from aquatic to terrestrial, and this was accomplished by amphibians. The fossil amphibians are rare in respect to other amphibians.

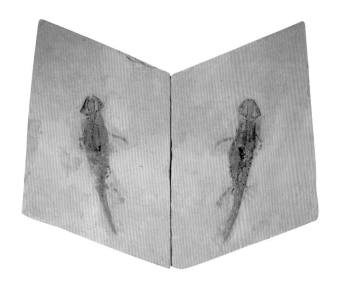
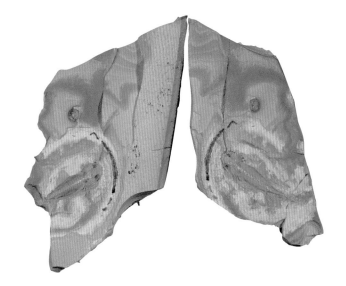
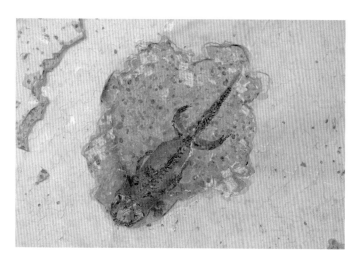
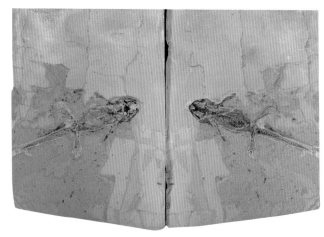

热河螈 *Jeholotriton* sp.

鱼 类 (Fishes)

鱼类是最低等的脊椎动物，用鳃呼吸，营水生生活。
Pisces is the lowest vertebrate. It breathes with its gills and lives in the water.

中华弓鳍鱼 (*Sinornithosaurus* sp.)

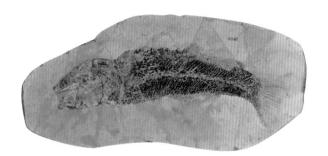
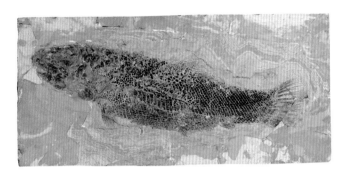

中华弓鳍鱼 *Sinornithosaurus* sp.

鲟 (Sturgeon)

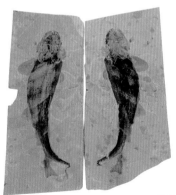

潘氏北票鲟 *Peipiaosteus pani*

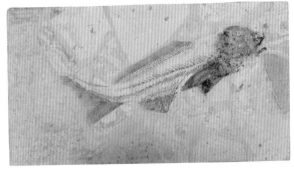

长背鳍燕鲟 *Yanosteus longidirsalis*

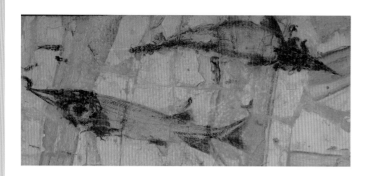 

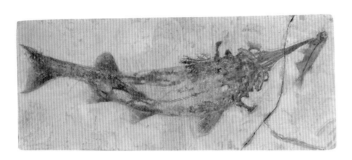 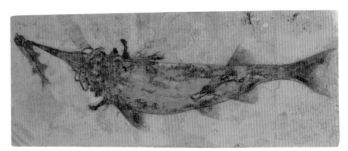

刘氏原白鲟 *Protopsephurus liu*

狼鳍鱼 (*Lycopter*)

狼鳍鱼是中生代后期东亚地区特有的淡水鱼类,广布于西伯利亚、蒙古、朝鲜和我国北部水域,是我国发现的最早的真骨鱼类。是热河生物群的化石代表之一。

Lycopter is a typical freshwater fish in late Mesozoic, they widely distributed in water areas in Siberia, Mongolia, Korea and North China, it is the earliest Teleostei found in our country, and it is the representation of Rehe biome fossils.

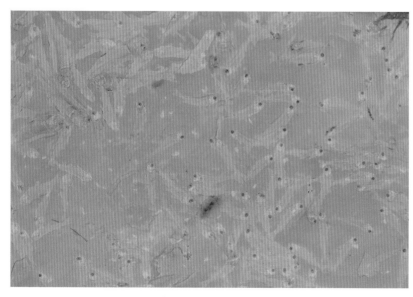 

狼鳍鱼 *Lycopter*

无脊椎动物 (Invertebrates)

虾类 (Shrimps)

虾是一类杂食性的甲壳动物。到目前为止,世界上发现的淡水虾化石还很少。

Shrimp is a kind of polyphagia crustacean. There have been only a few fossil fresh-water shrimps up to now.

 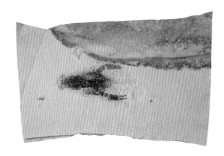

奇异环足虾 *Cricoidoscelosus aethus*

蜘蛛类 (Araneids)

蜘蛛是一种节肢动物,腹部圆而柔软,与头胸部接触处细小,钳角螯状,脚须简单。

Araneid belongs to a kind of arthropods. His belly is round and soft. The conjunction between head and breast is slim. The angles are like pincers and the foot beard is simple too.

 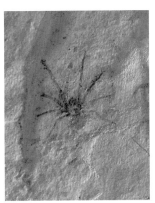 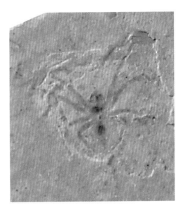 

蜘蛛 Araneids

昆虫类 (Insects)

昆虫是动物界中最大的一个纲。昆虫化石具有种类繁多、数量巨大、分布广泛、演化迅速的特点，是进行区域性地层划分和对比的化石材料。

Insecta is the largest class among the animal kingdom. Their fossils are very abundant, distribute widely and evolve rapidly, so they are good materials to distinguish and compare stratigraphies in areal geology.

三尾拟浮游 *Ephemeropsis trisetalis*

胡氏辽蝉 *Liaocossus hui*

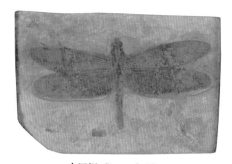

黑山沟衍蜓 *Aeschnidium heishankowense*　　　　中国蜓 *Sinaeschnidia* sp.

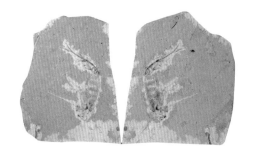

优鸣螽 *Habrohagla*

植 物 (Plants)

植物化石是研究地球生命出现和生物演化的重要组成部分,也是研究地球植被演化发展的重要证据。

Fossils plant are important portions in studying lives on earth and their evolution. They are also a basilic evidence in studying evolution and development of earthly vegetation .

辽宁古果 (*Archaefructus liaoningensis*)

辽宁古果是迄今为止世界上发现的最早的被子植物之一。草本水生,叶多次分裂为线状;果枝上螺旋着生有豆荚形的蓇葖果内含有2~4枚种子。

Archaefructus liaoningensis is one of the oldest angiosperm already found in the world. It is aquicolous herbage, its leaves are in linear shape after many times dissected. Reproductive axes bear helically arranged follicles (fruits), and there are 2-4 seeds in the fruit.

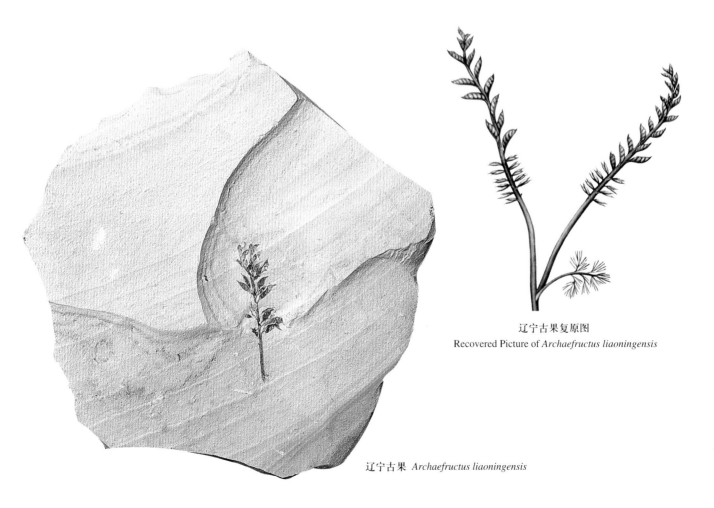

辽宁古果复原图
Recovered Picture of *Archaefructus liaoningensis*

辽宁古果 *Archaefructus liaoningensis*

中华古果 (*Archaefructus sinensis*)

中华古果是在辽西发现的第二种被子植物,它的营养叶与辽宁古果相同,但豆荚中含有 8~12 枚种子。

Archaefructus sinensis is the second prima angiosperm found in western Liaoning China, it has the same leaves as Archaefructus liaoningensis, but there are 8-12 seeds in the follicle.

中华古果 *Archaefructus sinensis*

中华古果复原图
Recovered Picture of *Archaefructus sinensis*

其他植物 (Other Plants)

瓣轮叶 *Lobatannularia*

轮叶 *Annulari*

枝脉蕨 *Cladophelibis*

似木贼 *Equisetitie*

新芦木 *Neoclamites*

锥叶蕨 *Coniopteris*

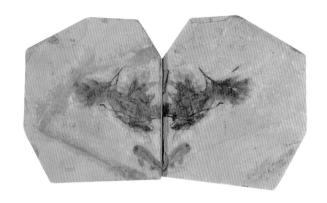
热河似荫地蕨 *Botrychites reheensis*

陈氏似麻黄 *Ephedrites chenii*

苏铁杉 *Podozamites*

薄氏辽宁枝 *Liaoningocladus boii*

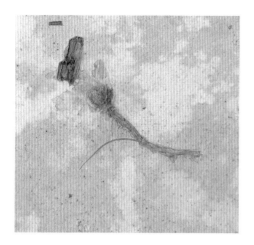
美丽威廉姆逊 *Wiliamsonia bella*

松型枝 *Pityocldus*　　　薄果穗 *Leptostrobus*

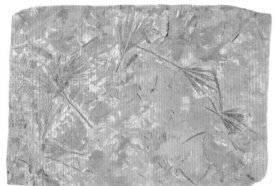

东北拜拉 *Baiera manchurica*

海尔松型球果 *Pterostrobus heer*

侧羽叶 *Pterophyllum*

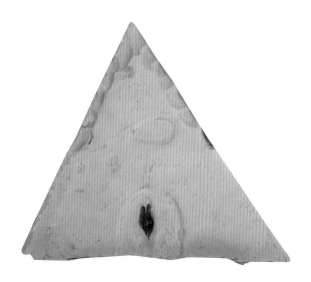

唇形裂鳞果 *Schizolepis chilitica*

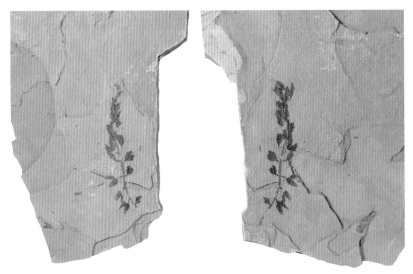

唇形裂鳞果 *Schizolepis chilitica*

毛羽叶 *Ptilophyllum*

紫荆 *Cercis linn*

色木槭 *Acermono maxim*

槭树 *Carpinus*

红杉 *Sequoia*

栗树 *Cadtanea mollissima*

枞型枝 *Elatocladus*

栎树 *Quercus*

赤杨 *Alnus*

模式标本 (Type Specimens)

孟氏丽昼蜓 (*Abrohemeroscopus mengi*)

孟氏丽昼蜓与昼蜓科的其他化石相比较，在演化上更为原始。主要体现在以下几个方面：后翅臀套较小，仅有 6 至 7 个翅室，径增脉（Rspl）缺；后翅 CuAa 脉弯曲，带有 5 个明显的后分支；前翅 MP 脉短，终止于翅后缘近翅结处；翅痣下有一个明显的支脉；后翅 CuAa 和 MP 域基部较窄，在三角室下方仅有 1 排翅室。

孟氏丽昼蜓的发现，对于九佛堂组的时代归属早白垩世提供了新的证据。

Compared with others fossils of Hemeroscopid, *Abrohemeroscopus mengi* have a more original process of evolution in following aspects: hindwing and loop is smaller, with only 6 ~ 7 cells ; Rspl is absent; the windwing vein CuAa is curved and has five distinct posterior branches; the forewing MP shortened,reaches the posterior wing margin slightly beyond the level of the nodus;pterostigmata more distinctly braced;the hindwing area between MP and CuAa is narrow,with only one row of cells near the discoidal triangle.

The discovery of *Abrohemeroscopus mengi* has provided new proofs for the opinion that Jiufotang Formation was early Cretaceous in age.

孟氏丽昼蜓 *Abrohemeroscopus mengi*

郝氏中国鸟龙 (*Sinornithosaurus haoiana*)

模式标本。其鉴定特征是：前颌骨主体部分相对较高，其长仅稍大于其高，前颌骨角大，前颌骨上突与鼻突均很长，上颌骨不参与外鼻孔的构成，上颌骨窗相对较小且为圆形，方颧骨上升突明显长于颧骨突，齿骨长高之比小；耻骨柄前后方向的宽度小于宽臼宽度模式标本。

Model specimen. This new specimen differs from S. *millenii* in that: the main body of the premaxillary is higher, its length being slightly longer than its height; the anterior margin of the premaxilla is vertical;the maxillary process of the premaxillary is very long; the maxilla is separated from external naris;maxillary fenestra is circular and relatively small;the ascending process of quadratojugal is remarkably longer than the jugal process; the ratio of the dentary length/height is distinctly small;and the pubic peduncle of ilium is longitudinally narrower than acetabulum and so on.

中国鸟龙复原图
Recovered Picture of *Sinornithosaurus*

郝氏中国鸟龙 *Sinornithosaurus haoiana*

孟氏大连蟾 (*Dalianbatrachus mengi*)

该标本为模式标本，保存极为完美，是迄今为止唯一一件具有完整皮肤印痕的蛙类化石。其鉴定特征是：头骨大，且宽大于长，上颌骨上具密集的梳状细齿，额顶骨愈合。肩带弧胸型，椎体后凹型，荐前椎9枚，前3枚躯椎具有短的肋骨。荐椎横突宽阔，呈大的扇形。尾杆骨长于荐前椎总长度。前肢粗短，后肢细长。胫腓骨与股骨等长，跗节长小于胫腓骨的一半。

This specimen is a model with perfect preservation and it's the only discovered frog fossil with complete skin moulage. The assigned features are that it has a large skull and the width of skull is larger than the length, there are dense tiny teeth on maxillary and the frontal-parietal is concrescisence. Pelvic gridle is arch shape and the spine is concave at the back. There are 9 presacrals vertebrae and the first 3 dorsal vertebrae have short ribs. The transversal process of sacrals is broad in large sector shape. The length of urostyle is larger than the total length of presacrals. The front limbs are short and thick, and the rear limbs are long and thin. The tibiofibula has the same length of thighbone and the tarsus is shorter than half of the tibiofibula.

孟氏大连蟾复原图
Recovered Picture of *Dalianbatrachus mengi*

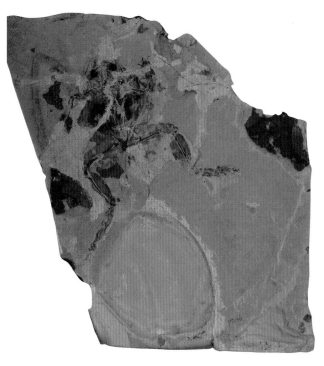 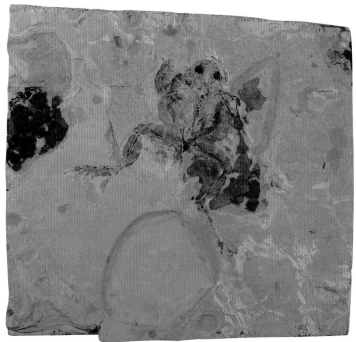

孟氏大连蟾 *Dalianbatrachus mengi*

朝阳辽西龙 (*Liaoxisaurus chaoyangensis*)

朝阳辽西龙时发现于辽宁省西部朝阳地区下白垩统九佛堂组的一具有头骨的不完整离龙类骨架。其特征是：下颌缝合部短，小于下颌长度的20%，牙齿齿槽近似正方形；吻部相对较短，占头骨总长的49.8%。

Liaoxisaurus chaoyangensis, an incomplete skeleton with a skull, comes from the Early Cretaceous Jiufotang Formation of western liaoning. It is distinguished from other derived Choristodera in that the short mandibular symphysis, less than 20% the length of the lower jaw, and the nearly square tooth alveolus. Liaoxisaurus chaoyangensis has a relatively short rostrum, which occupies 49.8% of the whole skull length.

朝阳辽西龙 *Liaoxisaurus chaoyangensis*

大连马 (*Equus dalianensis*)

大连马颊齿的大小、构造均与现生的普氏野马相似，但其大小不亚于欧洲同时代的大型野马。

Though it is close to Equus przewalskyi in size of upper and lower cheek teeth, *E. dalianensis* is quite large horse nothing less than which lived in Europe at the same time.

大连马 *Equus dalianensis*

蒋氏网形蛋 (*Dictyoolithus jiangi oosp*)

蛋化石扁卵圆形或扁圆形。蛋壳有不规则的基本结构单元重叠一起构成，排列松散。气孔道形状不规则。

Eggs, flat oval or oblate in shape. Eggshell layer of superimposed basic structural units loosely. Pore canals irregular.

海洋——生命的摇篮
Ocean: The Cradle of Life

　　海洋，辽阔而深邃，美丽而富饶。

　　海洋，是丰富多彩的世界，是生命的摇篮。从只有一个细胞的原生动物，到 100 多吨重的蓝鲸；从炎热的赤道到冰冷的两极；从蔚蓝的海面，到黑暗的大洋深处，繁衍生息着近 20 万种海洋生物。

The ocean is broad and deep, beautiful and rich.

The ocean is a colorful world which is the cradle of the life. There are nearly 200 thousand species of halobios living there: from the protozoan with only one cell to the blue whale with a weight of more than 100 tons; from the hot equator to the cold polar areas; and from the blue offing to the dark deep ocean.

海洋哺乳动物 (Marine Mammals)

海洋哺乳动物指哺乳类中适于海栖的特殊类群,简称海兽。一般包括鲸类、海牛类、鳍脚类、海獭及北极熊。海兽保持着哺乳类的共同特点,如胎生、哺乳、体温恒定和用肺呼吸等。

Marine mammals refer to the special group suitable for residence in the ocean. This group mainly includes cetacea, sirenian, pinniped, sea otter and polar bear. The marine mammals still have the common features of mammals such as viviparity, lactation, fixed body temperature and lung-breathing.

海洋哺乳动物展厅
The Hall of Marine Mammals

海洋巨兽－鲸 (Huge Animals in the Ocean-Whale)

鲸类的分布从热带海域到两极冰洋，少数生活在淡水河流中。现存78种，是地球上现存最庞大的动物。鲸体表光滑不被毛，皮下脂肪很厚；外鼻孔1或2个，位于头顶，俗称喷气孔；用肺呼吸。最大的蓝鲸可达31米长，重约130吨，而最小的海豚却不到2米长。现存鲸类有须鲸和齿鲸2个亚目。

Whale distributes from tropical to polar oceans while a minority of them lives in river. There are totally 78 species at present; it is the largest animal alive on the earth. The skin of whales are smooth and with thick fat underneath. They have 1 or 2 outer nostrils which are usually called as fumarole at the top of the head; the whales breathe with lungs. The largest blue whale can be as long as 31 meters and as heavy as 130 tons, however the smallest dolphin is less than 2 meters long. Present cetacean includes two suborder: mysticeti and odontoceti.

巨鲸展厅 The Hall of Whale

海洋：生命的摇篮

鲸须 (Baleen)

须鲸口中无齿，上腭两侧各生1列鲸须向下垂入口腔。鲸须由角质构成，每侧有须板130～450片，为滤食器官。须鲸成体大者30余米，共有4科11种，中国海域记录有3科8种。

Baleen whale does not have teeth, there is a line of baleen on each side of the upper jaw dropping into the mouth. The baleen is composed by horn, there are 130 -450 baleen boards on each side serving as food filtering organ.An adult baleen whale may be as long as more than 30 meters, totally 4 families 11 species. There are 3 families 8 species in Chinese maritime space recorded.

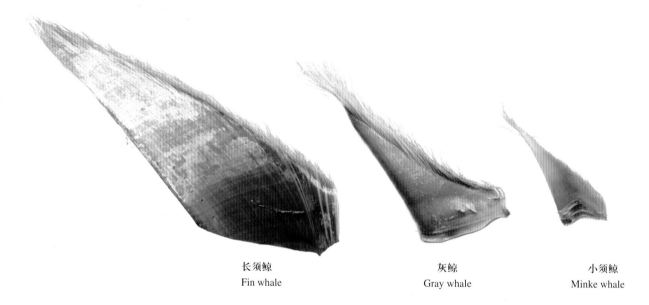

长须鲸　　　　　　　灰鲸　　　　　　　小须鲸
Fin whale　　　　　Gray whale　　　　Minke whale

鲸须比较图
鲸须的片数、颜色、形状和长度，是分类的依据之一。

Baleen Comparison
The number, color, shape and length of baleens are the main base for classification.

 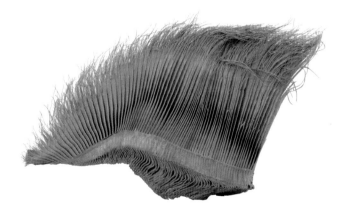

小须鲸须板内侧显示较粗的须毛　　　　　　　　　　小须鲸须板外侧显示较短的须板
Inner side of the baleen board of Minke whale Displays thick baleen　　Outer side of the baleen board of Minke whale Displays short baleen board

须鲸类两种摄食方法
Two ways of food taking

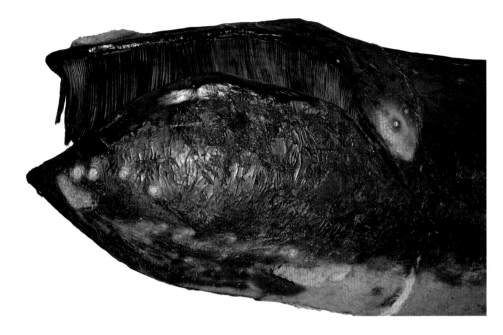

撇水型
露脊鲸等微微张开双唇,缓缓游动,海水流进嘴里,经过须板筛滤后,从两侧流出,留下食物。

Water Filtering
Right whale opens its two lips slightly and swims slowly, when sea water flows into its mouth and filtered by baleen board, the water will flow from two sides and leave the food.

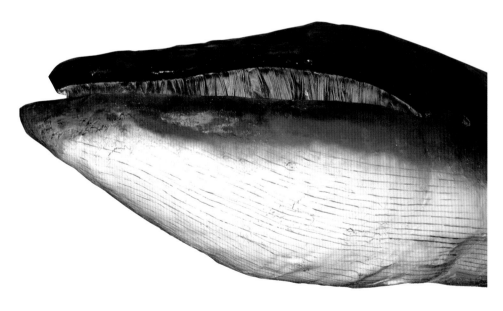

吞水型
长须鲸等具有喉褶的鲸,一口吞下大量海水,然后闭上嘴,收缩喉褶,将海水挤出去,留下食物。

Water Swallowing
Whales with throat grooves such as fin whale usually swallow a large amount of sea water and close the mouth, then contract the throats and squeeze the sea water out, leaving food there.

矮胖的黑露脊鲸 (Northern Right Whale)

黑露脊鲸 *Eubalaena glacialis* (Müller, 1776)

保护等级：中国：Ⅱ；CITES：附录Ⅰ；IUCN：濒危（EN）。

黑露脊鲸隶属露脊鲸科露脊鲸属，身体短粗，没有背鳍，头大，且覆有角质瘤。体呈黑褐色，腹面颜色较浅，有云状或条纹状斑块，鳍肢和尾鳍均呈黑色。黑露脊鲸行动迟缓，喜栖息于水的上层，把整个背部露出水面，故名露脊鲸。

该标本为1977年在黄海北部海域捕获，雌性，全长17.10米，体重约66.7吨，是我国保存的鲸类动物标本中体重最大的。

Northern Right Whale
Protection Class: China: Ⅱ; CITES: Appendix Ⅰ; IUCN: Endangered (EN).
Northern Right Whale belongs to Eubalaena of Balaenidae, the body is short and fat, there is no dorsal fin, its head is large covered with cutin burl. The body is in black brown color, the color of the belly is light with spot in cloud or stripe shape, the limb and tail are all black. The Northern Right Whale is slow in movement and loves to stay in the upper layer in the water to expose its back.
This specimen was caught in North Yellow Sea in 1977, female, the totally length is 17.10m with a weight of 66.7 tons; it is the whale specimen with the largest weight in China.

黑露脊鲸体长一般13~17米，体重可达80~100吨。喷潮时雾柱高达6米，呈两支，各向两侧倾斜呈V型。

Generally the length of Northern Right Whale is 13-17m with a weight of 80-100 tons. It can spray out two 6-meter water columns in V shape towards two sides.

海洋：生命的摇篮

黑露脊鲸夏季在北太平洋及北大西洋高纬度索饵，冬季向低纬度越冬，12月至翌年3月间进入中国黄海、东海及南海东部海域。

In summer, the Northern Right Whale seek foods in high latitude areas in the north Pacific and north Atlantic, and it travels to lower latitude area in winter, from December to next March it enters the Yellow Sea, the east China Sea and South China Sea.

长途迁徙的灰鲸 (Long-distance Traveling Gray Whale)

灰鲸 *Eschrichtius robustus* (Lilljeborg, 1861)

保护等级：中国：II；CITES：附录 I。

灰鲸隶属灰鲸科灰鲸属，体粗短呈纺锤型，头部较短。头部表皮上有如同被针刺的针眼分布。体色为暗灰色。

该标本是1996年12月在大连庄河沿海近岸搁浅死亡的，雌性，体长11.95米。

Gray whale

Protection Class: China: II；CITES: Appendix I.

Gray whale belongs to Eschrichtius genus of Eschrichtiidae family, the body is short and thick in spindle shape, and the head is relatively short. There are holes in the head cuticle like pierced by needles. Its body color is dark grey.

This specimen died on ground in December 1996 at the costal area in Zhuanghe, Dalian, female, the length is 11.95 meters.

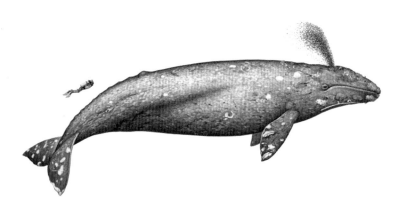

灰鲸体长一般12～13米，体重可达35吨。树丛状喷气柱通常高达3到4.5米。

The body length of gray whale is 12-13meters, the weight can reach 35 tons. The whiffing column in bosk shape can be as high as 3 to 4.5 meters.

灰鲸一般经日本海西了回游,有少量成年鲸沿朝鲜西海岸北上至黄海北部,部分沿东海中国沿岸继续南下。

Grey whale generally swims along the western coast of the Japan Sea, a small number of adult whales swim along the western coast of Korea to the northern Yellow Sea. Some of them swim along the East China Sea coast to the south.

苗条匀称的长须鲸 (Slim and Symmetry Fin Whale)

长须鲸 *Balaenoptera physalus* (Linnaeus,1758),

保护等级：中国：Ⅱ；CITES：附录Ⅰ；IUCN：濒危（EN）。

长须鲸隶属须鲸科须鲸属，体稍细长，呈纺锤形。背鳍呈镰状，其后缘凹进。体腹面有54～78条褶沟，由下颌前部纵向延伸至脐后。体色左侧比右侧色浓。多成群游动，游泳速度可超过每小时30公里。

该标本是1959年在黄海北部捕获的，雌性，体长18.40米，体重约34.7吨。

Fin whale

Protection Class: China: Ⅱ；CITES: Appendix Ⅰ；IUCN: Endangered (EN).

Fin whale belongs to Balaenoptera genus of Balaenopteridae family, its body is long and in spindle shape. The dorsal fin is in sickle shape with recessive back edge. There are 54-78 throat grooves in the belly from the frontal part of mandible to the navel. The color of the left part of the body is denser than the right part. They usually swim in groups with a speed over 30 km per hour.

This specimen is taken in 1959 at north Yellow Sea, female, the length is 18.40m with a weight of 34.7 tons.

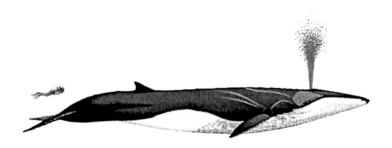

长须鲸体长仅次于蓝鲸，可达27米，体重可达75吨。
呼吸时喷出的雾柱细高，像倒置的圆锥状，上部顶端散开。大潜水后的第一个雾柱高达8～10米

The body length of fin whale is the second to blue whale, the length can reach 27 meters and the weight can reach 75 tons.

The fog column sprayed out is thin and tall in reverse taper shape, the top is fanning out. The first fog column after long diving can reach 8 to 10 meters.

海洋：生命的摇篮

长须鲸分布范围很广，但大部分生活在冷水区和南半球。中国南海、东海和黄渤海均有分布。

Fin whales live in vast areas, but most of them live in cold water area and the south hemisphere. They can be found in the South China Sea, the East China Sea, the Yellow Sea and Bohai Sea of China.

数量众多的小须鲸 (Numerous Minke Whales)

小须鲸 *Balaenoptyera acutorostrata* (Lacepede, 1804)

保护等级：中国：Ⅱ；CITES：附录Ⅰ

小须鲸隶属须鲸科须鲸属，体粗短，呈纺锤形。腹部褶沟50~72条。鳍肢外侧中央部分有1条宽约20~35厘米的白色横带。

该标本是2000年4月在大连沿海搁浅死亡的，雌性。

Minke whale

Protection Class: China: Ⅱ；CITES: Appendix Ⅰ

Minke whale belongs to Balaenopteragenus of Balaenopteridae family, the body is thick and short in spindle shape. There are 50-72 throat grooves in the belly. There is one 20-35 cm long white stripes on the central outer flipper.
This specimen was found dead ashore in April 2000, female.

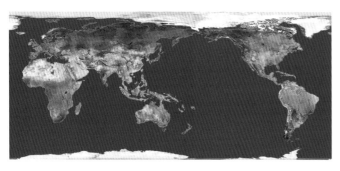

小须鲸体长不超过10米，一般5~10吨。
呼吸时喷出的雾柱细而稀薄，高达1.5~2米，消失得快，不易观察。

The length of Minke whale generally does not exceed 10 meters, generally 5-10 tons.
The fog sprayed out when breathing is thin and rare with a height of 1.5 to 2 meters, it quickly disappears and cannot be easily observed.

小须鲸骨骼
Skeleton of Minke whale

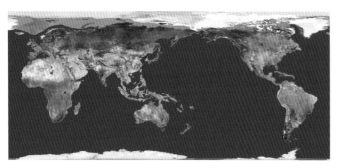

小须鲸是最小的须鲸，其踪迹几乎遍布全球海域。

Minke whale is the smallest baleen whale, it can be found almost any ocean on the earth.

鲸齿 (Whale Teeth)

齿鲸类体长大者可达 20 多米，背鳍或有或无，尾部比例较须鲸类大。1 个外呼吸孔。口中有齿，呈圆锥状或纵扁形，是分类的主要依据之一。齿鲸亚目共有 9 科 67 种，中国水域记录有 7 科 28 种。

Large toothed whales can be as long as more than 20 meters, some have dorsal fin and some don't, the proportion of the tail is larger than baleen whale. It has one outer breathing hole. There are teeth in its mouth in taper shape or flat, it is one of the main principles in classification. There are 9 families 67 species in odontoceti suborder, there are 7 families 28 species recorded in Chinese water area.

鲸齿   Whale Teeth

虎鲸上、下颌各有圆锥形的利齿 20～26 枚，长约 8～13 厘米。

Killer whale has 20-26 sharp teeth in taper shape both at the upper and bottom jaws, the length is 8-13 cm.

伪虎鲸上、下颌各有长达 8 厘米的长牙 16～22 枚。

False Killer Whale has 16-22 sharp teeth in taper shape both at the upper and bottom jaws, the length is 8cm.

真海豚的上、下颌各有小而尖的齿 82～108 枚

Common Dolphin has 82-108 tiny and sharp teeth in both upper and bottom jaws

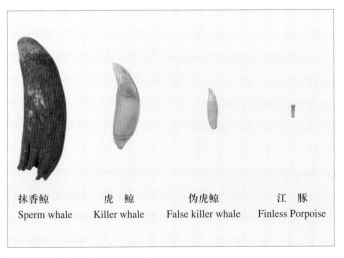

抹香鲸　虎鲸　伪虎鲸　江豚
Sperm whale　Killer whale　False killer whale　Finless Porpoise

牙齿比较图 Teeth Comparison Picture

鲸的牙齿生长是分层的，每长一岁就增加一层。

The teeth of whales are in layers, one more layer will grow after one year.

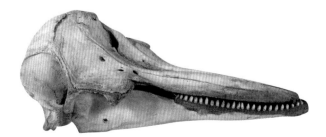

宽吻海豚的上、下颌各有结实的齿 36～52 枚

Bottlenose has 36-52 strong teeth in both upper and bottom jaws.

潜水冠军抹香鲸 (Diving Champion Sperm Whale)

抹香鲸 *Physeter macrocephalus* (Linnaeus, 1758)
保护等级：中国：Ⅱ；CITES：附录Ⅰ

抹香鲸隶属抹香鲸科抹香鲸属，为齿鲸中的巨无霸。头部特别巨大，呼吸孔位于头顶前端偏左侧。体色多为蓝黑色或黑褐色。

抹香鲸潜水本领很强，最大记录达2,200米。大潜水时可在水中停留55分钟。

该标本是1993年2月在东港市前阳镇沿海近岸搁浅死亡的，体长17.5米，重50余吨。

Sperm whale
Protection Class: China: Ⅱ；CITES: Appendix Ⅰ
Sperm whale belongs to the Physeter genus of Physeteridea family, it is the biggest among toothed whales. It has a huge head and the breathing hole is the front of the head top (slightly left). The body color is mainly blue black or black brown.

Sperm whale has a very strong diving skill; the maximum record is 2200 meters. It can stay in the water for 55 minutes when diving.

This specimen is dead on ground in February 1993 in the coastal area of Qianyang Country, Donggang City; its length is 17.5 meters with a weight of more than 50 tons.

抹香鲸雄性最大20米，最大体重约57吨。
喷出的雾柱向前偏左同水面呈45度角，可高达5米。

The largest size of male sperm whale is 20 meters, and the maximum weight is 57 tons. The fog column sprayed out is in 45 degree angle with the water surface, the length can reach 5 meters.

抹香鲸主要鲸群见于南北纬 70 度间的热带、亚热带水域。中国以东海、南海较多。
Sperm whales mainly live in the tropical and semi-tropical ocean area between 70 degree latitude of south and north hemisphere. The East China Sea and the south China Sea has some sperm whales.

海中杀手虎鲸 (Killer in the Ocean: Killer Whale)

虎鲸 Orcinus orca（Linnaeus,1758）

保护等级：中国：Ⅱ；CITES：附录Ⅱ。

虎鲸隶属领航鲸科虎鲸属，体呈纺锤形。背鳍高大，呈三角形，位于身体中部稍前方。背部黑或灰黑色，腹面白色界限十分鲜明。

虎鲸常成群追食海豹、海狗、海豚等海兽，甚至追击大型的须鲸。

Killer whale

Protection Class: China: Ⅱ；CITES: Appendix Ⅱ.

Killer whale belongs to the Orcinus of Delphinidea.It is in spindle shape with triangle high and large dorsal fin, the dorsal fin is in the upper-middle area of the body. The back is black or grey black, the white boundary of the belly is very obvious.

Killer whales usually prey on seals, fur seals and dolphins, etc, and they even prey on large baleen whale.

虎鲸为大洋种，没有固定栖息场所，主要追逐猎物移动，从潮间带到外海都可以出现。

Killer whale lives in ocean without fixed residence, it mainly preys on moving quarries, they can appear from tidal sea to open ocean.

体长雌性可达8.5米，重约7.5吨；雄性可达9.8米，重约10吨。

The length of female killer whale can reach 8.5 meters with a weight of 7.5 tons; the length of the male can reach 9.8 meters with a weight of about 10 tons.

"集体自杀"的伪虎鲸 (False Killer Whales "Suicide in Groups")

伪虎鲸 Pseudorca crassidens (Owen, 1846)

保护等级：中国：Ⅱ；CITES：附录Ⅱ。

伪虎鲸隶属领航鲸科伪虎鲸属，身体匀称细长。眼小。1个鼻孔。

伪虎鲸常结成十余头或百余头甚至千余头的大群。

False killer whale

Protection Class: China: Ⅱ ; CITES: Appendix Ⅱ.

False killer whale belongs to Pseudorca of Delphinidea, its body is symmetric and slim. The eyes are small and it only has one nostril.

False Killer Whales usually stay in groups composed by more than ten whales or even hundreds or thousands.

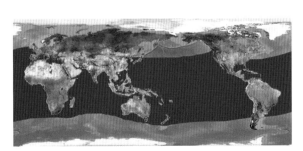

伪虎鲸数量并不多，但分布范围很广，世界各海洋均有。中国沿海各海区均有记载。

The false killer whales are not large in number but they are widely distributed in each ocean of the world. They are recorded at each sea of China.

体长雌性可达5米，雄性可达6米，重约2吨。

The length of female false killer whale can reach 5 meters, and male 6 meters with a weight of about 2 tons.

神秘的日本喙鲸 (Mysterious Ginkgo-toothed Beaked Whale)

日本喙鲸 *Mesoplodon ginkgodens Nishiwakietkamiya* (Nishiwaki and Kamiya, 1958)

保护等级：中国：Ⅱ；CITES：附录Ⅰ

日本喙鲸隶属喙鲸科喙鲸属，身体细长而侧扁。上颌无齿，仅下颌具1对扁平而大的齿。体背部呈蓝黑色，腹面较浅。体下侧及腹面均具有不规则的灰白色斑点。

该标本是1980年8月13日在辽宁庄河搁浅的雌鲸，体长4.05米，体重748公斤，为黄海海域初次记录。

Ginkgo-toothed Beaked Whale
Protection Class: China: Ⅱ ; CITES: Appendix Ⅰ
Ginkgo-toothed beaked whale belongs to Mesoplodon of Ziphiidea, the body is slim and flat in two sides. There are no teeth on upper jaw, and there is only one pair of large flat teeth in the bottom jaw. The back is blue black and the belly surface is in light color. There are irregular spots under the body and on the belly.
The specimen is a female whale taken in Zhuanghe, Liaoning Province in Aug 13th, 1980, the body length is 4.05 meters with a weight of 748 kg, and it is the first record of its appearance in Yellow Sea.

捕获记录最大体长雌鲸4.9米，雄鲸1.8米。
The largest captured female whale has a length of 4.9 meters and male 1.8 meters.

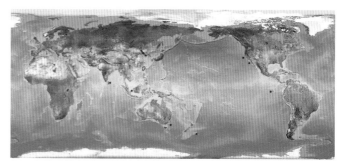

深蓝色区域为已知分布范围，红色圆点为目击或搁浅地点。
分布仅见于北太平洋西部海域，首次发现于日本九州近海，中国台湾海域仅有数次记录。

Areas in deep blue are areas already known, red spots signify on-looking areas or stuck areas.
It distributes only in the western Pacific Ocean, it is first found in the offing of Kyushu of Japan, In China, there are only records for several time in Taiwan .

中国特有的白鱀豚 (Yangtze River Dolphin: Peculiar in China)

白鱀豚 *Lipotes vexillifer* (Miller, 1918)

保护等级：中国：Ⅰ；CITES：附录Ⅰ；IUCN：极危（CR）

白鱀豚属白鱀豚科白鱀豚属，中国特产。头尾长 1.5～2.5 米，体重 100～200 公斤。喙长约 30 厘米。背面浅蓝灰色，腹面白色。栖息于淡水江河中。现有数量仅百头左右。

该标本 1966 年 3 月 3 日采于南京长江燕子矶，体重 86.5 公斤。

Baiji, or Yangtze river dolphin

Protection Class: China: Ⅰ; CITES: Appendix Ⅰ; IUCN: Extremely Endangered (CR)

Baiji, or Yangtze river dolphin belongs to Lipotes of Lipotidae, it is peculiar to China. Its length from head to tail is 1.5-2.5meters with a weight of 100-200kg. The length of its beak is 30 cm. Its back is in light blue grey color and its belly is white. It inhabits in freshwater rivers, totally there are only about 100 in number.

This specimen is picked at Yanziji, Yangtze River, Nanjing on March 3, 1966, with a weight of 86.5kg.

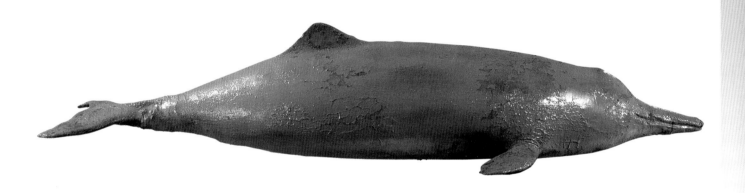

白鱀豚吻突狭长，上、下颌前端均略向上翘，有齿 60～76 枚

The beak of Yangtze River Dolphin is narrow and long, the frontal parts of upper and bottom jaws are slightly rising, there are about 60-76 teeth.

海豚大家族 (A Large Dolphin Family)

海豚有 26 种，多数种类吻突明显，体长小于 4 米，流线型身体，尾鳍中央都有凹刻。

There are totally 26 species of dolphins, most of them have obvious beak, the body length is less than 4 meters in stream line shape, there are concaves in central flukes.

灰海豚 *Grampus griseus* (Cuvier, 1812)
保护等级：中国：Ⅱ；CITES：附录Ⅱ
分类：海豚科灰海豚属

Risso's Dolphin
Protection Class: China: Ⅱ；
CITES: Appendix Ⅱ
Classification: Grampus of Delphinidea

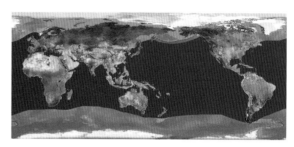

分布图 Distributing Chart

宽吻海豚 *Turiops truncatus* (Montagu, 1812)
保护等级：中国：Ⅱ；CITES：附录Ⅱ
分类：海豚科宽吻海豚属

Bottlenose Dolphin
Protection Class: China: Ⅱ；CITES: Appendix Ⅱ
Classification: Turiops of Delphinidea

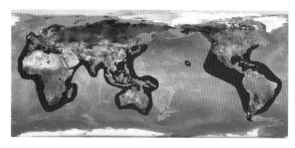

分布图 Distributing Chart

真海豚 *Delphinus delphis* (Linnaeus, 1758)
保护等级：中国：Ⅱ；CITES：附录Ⅱ
分类：海豚科真海豚属

Common Dolphin
Protection Class: China: Ⅱ；CITES: Appendix Ⅱ
Classification: Delphinus of Delphinidea

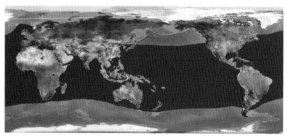

分布图 Distributing Chart

鲸的繁殖 (Reproduction of Whales)

像所有哺乳动物一样，幼鲸在母亲的子宫里成长，须鲸类要12个月，有些齿鲸甚至长达18个月。

Like all the other mammals, the baby whale grows up in the uterus of the mother, baleen whales need 12 months, toothed whales even need 18 months.

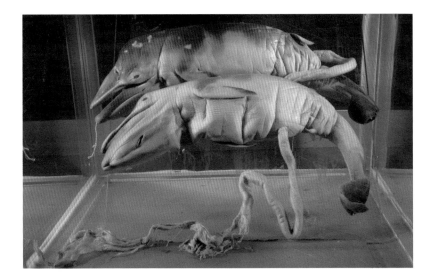

小须鲸双胞胎
由于野外研究鲸类动物的行为非常困难，即使解剖死鲸的尸体，见到多胞胎的情况也极为罕见。这是1974年捕到的一头小须鲸体内解剖到的一对双胞胎胎儿。

Twins of Minke whale
Since it is very difficult to do field research about the behaviors of the whales, the discovery of twins is very rare even when anatomizing the dead body of whales. These are the twins found when anatomizing the captured Minke whale in 1974.

宽吻海豚胎儿要在母体内发育6~8个月，出生后还要母亲照顾2年以上才独立生活。

Embryo of bottlenose dolphin needs to develop 6-8months within its mother's body, after born the mother has to take care of the baby for more than two years.

小黑露脊鲸每天要吃奶300升以上，体重每天能增加100公斤，体长每天增加3~4厘米。

Small northern right whale needs to take more than 300 liters of milk everyday to increase the weight by 100 kg per day, the body length increase 3-4 cm per day.

鲸的交流 (The Communication between Whales)

鲸类在水中是靠声音来辨别方向的。鲸没有声带，而且能够使肺部空气不断循环，所以它们不用像人类那样每隔几秒就得停下来吸气，鲸可以不间断地唱上几个小时。

Whales locate directions under water by sounds. Whales do not have vocal cords, the air in the lung can recycle continuously, so they do not need to stop to breathe like human beings, they can sing continuously for several hours.

在这里你可以听到鲸的歌声。
You can hear songs of whales here.

鲸的保护 (The Protection of Whales)

鲸是海洋生态系统的重要组成部分。由于人类的活动，使现在海洋里的鲸只是原来的5%～10%左右，而一些小型鲸类仍被人们捕猎，有些种类已经濒临灭绝。蓝鲸、黑露脊鲸这样的大型鲸类的数量可能永远不会回升了，尽管人们倍加努力，有些鲸可能还是会在几百年内从地球上消失。

Whale is an important part in the ocean eco-system. Due to human activities, at present the number of whales is only 5%-10% of the past, and some small whales are still hunted by human beings, some of them are nearly extinct. The number of large whale such as blue whale and northern right whale may not increase any more forever, although we are working hard to protect them, some whales will disappear in the world within next few centuries.

捕鲸炮带来的劫难
1868年挪威人斯文德·福因发明了捕鲸炮，这种捕鲸炮被架在蒸汽船的船头，可以发射尖端带炸药的标枪，一旦射中目标，标枪顶端的炸药数秒后会在体内爆炸，被射中的鲸根本就没有幸存的机会。

Disaster Brought about by Whale Fishery Cannon
In 1868 Norwegian Svend Foy invented whale fishery cannon which can be fixed at the fore to emit javelin with dynamite at the top, once the target is hit, the dynamite will explode in the body, there is no chance for the whale to survive.

灵巧活泼的鳍脚类 (Agile and Active Pinnipeds)

鳍脚类属食肉目动物，体呈纺锤形，体表密生紧贴于身体的短毛，尾短小，四肢变为鳍状，只在产仔、哺乳期才上陆。全球有3科34种，仅有2科5种进入中国海域。

Pinniped belongs to carnivorous animal and the body is in spindle shape, there is short hair on the surface of the body, the tail is short and small, the limbs are in fin shape, they come to the land only when they are littering or in the lactation. There are totally 3 families 34 genera in the world, only 2 families 5 species live in Chinese maritime space.

海狮和海豹的区别
The Differences between Sea Lion and Seal

	海狮 Sea Lion	海豹 Seal
前肢 Forelimb	强壮，长度超过身体的1/4 Strong, longer than a quarter of the body	短小，长度不及体长的1/4 Short, shorter than a quarter of the body
游泳 Swimming	主要靠前肢 Mainly depends on forelimb	靠后肢左右摆动 Left-right swing of rear limbs
后肢 Rear Limb	能够向前弯曲 Can bend forward	不能向前弯曲 Cannot bend forward
外耳 Outer Ear	有，长约5厘米 Have it, the length is about 5 cm	没有 None
皮毛 Fur	有小绒毛 Small fluff	没有小绒毛 No small fluff
陆地 Land	能支持身体在路上行走 It can walk on the land by supporting	只能靠身体向前蠕动 It can only wriggle forward by body movement

分布图
Distributing Chart

海狗 *Callorhinus ursinus* (Linnaeus, 1758)
保护等级：中国：Ⅱ；CITES：附录Ⅰ
海狗是一种海狮，主要分布在北太平洋及白令海和鄂霍次克海，中国黄海偶有发现。海狗雌雄异型明显，雄性体长约2.5米，体重约300公斤；雌性只有1.45米，体重仅60公斤。

Northern Fur Seal
Protection Class: China: Ⅱ；CITES: Appendix Ⅰ
Northern fur seal is a kind of sea lion, it mainly distributes in North Pacific Ocean, Bering Sea and Okhotsk, it is occasionally found in Huanghai Sea of China. There are great differences between male and female northern fur seal, the length of the male is about 2.50 meters with a weight of about 300 kg, and the length of the female is only 1.45m with a weight of 60 kg.

逗人喜爱的海豹 (Amusing Seals)

海豹主要分布在北冰洋、北太平洋、北大西洋和南极周围海域，全世界有19种。海豹的游速可达每小时27公里，一般可潜到100米。

Seals mainly distribute in the Arctic Ocean, North Pacific Ocean, North Atlantic Ocean and the area around the South Pole, there are totally 19 species in the world. The swimming speed of seals can reach 27km per hour and they can dive 100 meters under the water.

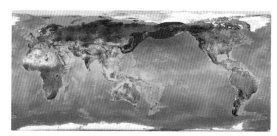

分布图
斑海豹广泛分布于西北太平洋白令海、鄂霍茨克海、日本海、黄海、渤海。

Distributing Chart
Larga seal widely distributes in northeastern Pacific Ocean Bering Sea, Okhotsk Sea, Japan Sea, Yellow Sea and Bohai Sea.

海洋：生命的摇篮

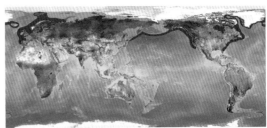

分布图
点斑海豹分布广泛，从鄂霍茨克海沿阿留申群岛到北美洲的沿海地区、北大西洋的美洲、欧洲沿海地区以及格陵兰岛沿海。

Distributing Chart
Larga seal widely distributes from Okhotsk along Aleutian Islands to the coastal areas in North America, America in North Atlantic, Coastal Europe and the Coastal area of Greenland.

点斑海豹 *Phoca vitulina* (Linnaeus, 1758)
点斑海豹在陆地上产仔，这是它和斑海豹的最大区别。标本1997年5月采于鄂霍茨克海，体长1.78米，体重100公斤。

Harbour seal
Harbour seal delivery in the land, and this is the most significant difference between harbour seal and the larga seal. The specimen is taken in Okhotsk in May 1997; its length is 1.78m with a weight of 100 kg.

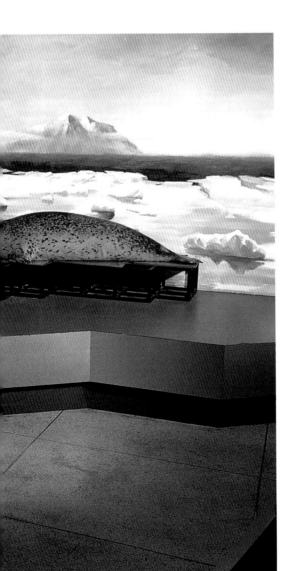

斑海豹幼仔正在褪毛
Young Larga Seal is molting

斑海豹 *Phoca largha* (Pallas, 1811)
保护等级：中国：Ⅱ；CITES：附录Ⅰ
斑海豹体长1.5～2.0米，雄性体重达150公斤。体呈纺锤形，头圆眼大，吻短而宽。成体灰黄色或深灰色，背部多斑点。初生仔被有白色绒毛。主要栖息于北半球高纬度海域。
斑海豹是唯一能在中国海域繁殖的鳍脚类动物。每年11月前后，进入渤海，至辽东湾。1～2月在浮冰上产仔，每胎1仔。

Largha seal
Protection Class: China: Ⅱ；CITES: Appendix Ⅰ
The length of larga seal is 1.5 to 2.0 meters, the male larga can be as heavy as 150kg. Its body is in spindle shape, the head is round and eyes are large, the beak is wide and short. The adult body is yellow or deep grey, there are many spots on its back, and the newly-borns are covered with white fluff. They mainly inhabit in high latitude areas in the north hemisphere.
Larga is the only pinniped that can reproduce in Chinese maritime space. Every year after November they enter Bohai Sea and reach East Liaoning Bay, then reproduce in the floating ice from January to February, they will give birth to one baby in each delivery.

贪食而聪明的海狮 (Gluttony but Clever Sea Lion)

海狮种类很多，有14种。除繁殖期外，没有固定的栖息场所。主要吃乌贼和鱼类。海狮十分聪明，平衡器官特别发达，所以经过训练后能够表演。

There are totally 14 kinds of sea lion. They do not have fixed residence except in the reproduction period. They mainly prey on cuttlefish and fishes. The sea lions are very clever and have fully developed balancing organs, so they can give performance after training.

南海狮 *Otaria flavescens* (Shaw,1800)
保护等级：CITES：附录 I
南海狮雄兽体长 2.5 米，重 350 公斤。两性异型。

South American Sea Lion
Protection Class: CITES: Appendix I
The length of male southern sea lion is 2.5 meters with a weight of 350 kg; the animals of different sex are different in body type.

加州海狮 *Zalophus californianus* (Lesson, 1828)
保护等级：CITES：附录 I
加州海狮雄兽体长 2.4 米，重 390 公斤。

California Sea Lion
Protection Class: CITES: Appendix I
The length of the male California sea lion is 2.4 meters with a weight of 390 kg.

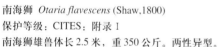

加州海狮
California Sea Lion

南海狮
South American Sea Lion

分布图
南海狮分布广泛，从秘鲁北部向南沿南美西海岸一直到东海岸巴西南部。
加州海狮的分布，从加拿大的温哥华沿北美西海岸一直到墨西哥中部。

Distributing Chart
Southern American sea lion widely distributes from the north Peru down along the South America coast to the eastern coast, i.e. the southern part of Brazil. California sea lion mainly distributes from the Vancouver, Canada along the western coast of North America to middle Mexico.

美人鱼儒艮 (Mermaid Dugong)

　　儒艮属海牛目，是一类体型粗大的水生哺乳动物，皮肤厚，多皱褶，体表几乎无毛，但唇周多硬的触须。无耳壳，眼很小，前肢鳍状，后肢已退化，栖息于浅海及河口。全球现存2科2属4种。成年儒艮体长可达3.3米，体重400公斤以上。成体暗棕色。

Dugong belongs to sirenian, it is a large marine mammal, the skin is very thick with some folds, there is almost no hair on the body surface; there are dense bristles on the muzzle. It does not have ear shell, the eyes are very small, the forelimbs are in fin shape and the rear limbs are degenerated, they inhabits in shallow seas and river influx. There are totally 2 famlies 2 genera 4 species in the world. Adult dugong can reach 3.3 meters long and at least 400 kg, the adult body is in dark brown color.

儒艮 *Dugong dugon* (Müller,1776)
保护等级：中国：Ⅰ；CITES：附录Ⅰ；IUCN：易危（VU）

Dugong
Protection Class: China: Ⅰ; CITES: Appendix Ⅰ; IUCN: Very Endangered (VU)

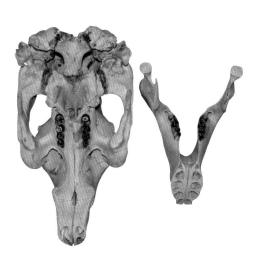

儒艮以扁平的臼齿磨碎海草
The flat cheek teeth suit to grind seaweed

分布图
儒艮广泛分布于印度洋沿岸及太平洋东南亚沿岸以及澳大利亚西部和北部沿海。我国见于台湾、广东、广西沿海浅海区域。

Distributing Area
Dugong widely distributes coastal areas of the Indian Ocean and Southeast Asian along the Pacific Ocean, and the west and north coastal area of Australia. It can be found in shallow water along coastal Taiwan, Guangdong and Guangxi.

海洋鱼类 (Ocean Fishes)

鱼类是指用鳃呼吸、用鳍游泳、多数体表生有鳞片的脊椎动物。世界上有鱼类25,000多种，是脊椎动物中最大的一个"家族"。个体大的鲸鲨有20米长，小的矮虾虎鱼，体长仅有0.75~1.15厘米。我国有鱼类3,000多种。

Fish refers to the vertebrate breathing with branchia, swimming with fins and the body is covered by squama. There are totally more than 25,000 species of fish; they are the largest "family" in the vertebrate. The size of large Whale Sharkcan reach 20 meters, and the small freckle-faceonly has a body length of 0.75-1.15cm. Our country has more than 3,000 species of fish.

大连自然博物馆馆藏
The Collection of Dalian Nature History Museum

软骨鱼展厅
The Hall of Selacean

海洋软骨鱼类 (Ocean Elasmobranch)

软骨鱼类没有鳔,骨骼为软骨性,没有真骨组织,适宜海洋生活。它们的口具上下颌,口中有牙齿。鼻孔1对。肉食性。卵生或卵胎生,现存约846种,我国产260种。可食用;制鱼肝油、鱼粉、皮革、工艺品;鱼鳍可加工鱼翅,鱼骨有治癌等功效。

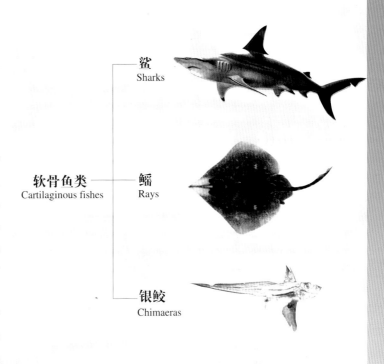

Elasmobranch does not have swim bladders and their bones are soft, there is no real bone organization, this guarantees they are suitable for living in the sea. They have upper and bottom jaws with teeth in the mouth. They have one pair of nostrils. They are flesh-eating animals, they are oviparous or ovoviviparous, and at present there are totally 846 species among which our country has 260 species. They are edible, and they can be used to produce fish liver oil, fish powder, leather and art craft; the fins can be processed into edible fin, and the bones have the function of curing cancer.

鲨 (Sharks)

鲨鱼是人类最为害怕的海洋鱼类,其视觉、嗅觉以及鼻子附近专门用来侦测电场的器官非常灵敏,一旦察觉到猎物的踪迹,便立刻以惊人的速度冲上前去,用足以咬啐骨头的力量咬住猎物。鲨鱼因其骨骼为软骨,故称为软骨鱼类。鲨类共有369种,我国产118种。

Shark is the most formidable sea fish to us human beings, it has very acute vision and smell, and the electric field sensing organs around the nose. Once they notice the trace of their quarry, they will rush towards in astonishing speed and bite the quarry with a force strong enough to snap apart the bones. The bones of sharks are soft, so they are called elasmobranch. There are totally 369 species of sharks, among which our country has 118 species.

软骨鱼展厅
Elasmobranch Exhibition Hall

鲨鱼的牙齿 Teeth of Sharks

在鲨鱼中，噬人鲨、鼠鲨、灰鲭鲨、大青鲨、锥齿鲨、鼬鲨及双髻鲨等凶猛鲨鱼，是人类最害怕的海洋杀手。鲨鱼的牙齿是皮肤的衍生物，从颌骨后面长出来，可以碌动更新。

Among all kinds of sharks, fierce sharks such as Man-eater shark, porbeagle, Mako shark, blue shark, ragged-tooth shark, Tiger shark and Mallet shark are the most formidable killers to human. The teeth of sharks are ramifications of their skin, the teeth grow from the rear part of the jaw, and the teeth can change regularly.

鲨鱼的牙齿
Teeth of Sharks

姥鲨的牙齿
Teeth of Bone Sharks

鲸鲨的牙齿
Teeth of Whale Shark

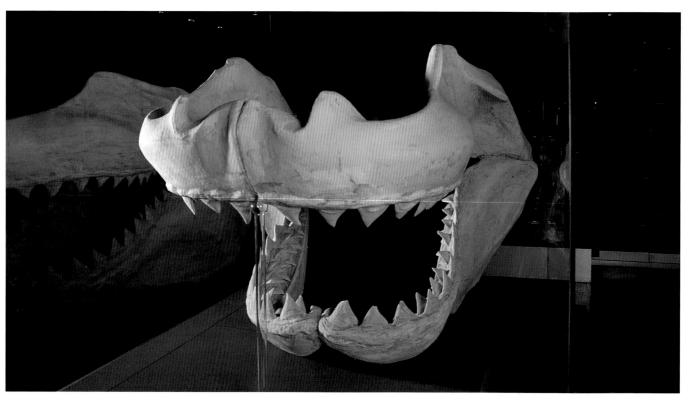

鲨鱼颌骨 Jaws of Sharks

鲸鲨 Whale Shark

　　鲸鲨是最大的鱼，体长可达18米，体重达40吨。口中长着成排的小齿，俗称"齿状突起"。游泳极为缓慢，它们一边游动，一边过滤水中的浮游动物。其为海洋上层大型鱼类，以浮游无脊椎动物为食，分布在南海、台湾海峡、东海、黄海等。

Whale Shark is the largest fish, its body length can reach 18 meters with a weight of 40 tons. There are rows of small denticles in its mouth, which is called "dentation". The swimming speed is very slow; they can swim while filtering the zooplanktons in the water. They are large fish in the upper layer of the ocean, they prey on pelagic invertebrate, and distribute in Nanhai Sea, Taiwan Straits, Donghai Sea and Huanghai Sea.

鲸鲨 *Hincodon typus* Smith. 1929
保护等级：CITES 附录 Ⅱ

Protection Class: CITES Class Ⅱ

姥鲨 Bone Shark

体长7~9米，体重3~5吨。头大，锥形。齿细小而多，圆锥状。为海洋上层大型鱼类。以浮游无脊椎动物为食，游动缓慢、性情温和，无危害。我国分布于南海、东海、黄海。

The body length is 7~9 meters with a weight of 3~5 tons. The head is large taper. The teeth are tiny and numerous in taper shape. It is a large fish in the upper layer of the ocean, and it preys on floating invertebrate, they swim very slowly with a mild temper and no harm. It can be found in Nanhai Sea, Donghai Sea and Huanghai Sea of our country.

姥鲨 *Cetorhinus maximus* (Günner 1765)
保护等级：CITES 附录 II

Protection Class: CITES Appendix II

其他的鲨 Other Sharks

路氏双髻鲨 *Sphyrna lewini* (Griffith en Smith)
为暖温性外海中大型鲨鱼。性凶猛，肉食性。分布于黄海北部。

Scalloped hammerhead is a large shark in warm open sea, it is very fierce flesh-eater and distributes in north Huanghai Sea.

狐形长尾鲨 *Alopias vulpinus* (Bonnaterre)
长尾鲨的尾巴很长，约占其全长之半。体黑褐色。为暖水性上中层大洋鱼类。

Bonnaterre has a very long tail, which is nearly half of its body, its body is black brown, and it is ocean fish in warm upper-middle layer.

噬人鲨 *Carcharodon carcharias* (Linnaeus)
为大洋暖水性游泳敏捷的鱼类。性凶猛，捕食各种大型水生动物。该标本为 1980 年 8 月 28 日采自渤海。

It is agile fish in warm ocean area, it is very fierce and mainly preys on large aguatic animal, and this specimen is taken in Bohai Sea on Aug 28, 1980.

居氏鼬鲨 *Galeocerdo cuvier* (Peron et Lesueur, 1822)
曾有过食人记录

There is a record for tiger shark of eating human beings.

灰鲭鲨 *Isurus glaucus* (Müller et Henle)
暖水性大洋表层种类。性凶猛。

Mako shark lives in the surface of warm ocean and its temper is very fierce.

黑印真鲨 *Carcharhinus menisorrah* (Müller et Henle)
为暖水性近海底层栖息中型次要经济鱼类。分布于黄海北部，少见。

Blackspot shark is minor economic fish living in the near bottom of the warm ocean. It mainly distributes in north Huanghai Sea and it is very rare.

扁头哈那鲨 *Notorhynchus platycephalus* (Tenore)

阴影绒毛鲨 *Cephaloscyllium umbratile* Jordan et Fowler.

白斑星鲨 *Mustelus manazo* Blecker

短吻角鲨 *Squalus brevirostris* Tanaka.

狭纹虎鲨 *Heterodontum zebre* (Gray)

鳐 (Rays)

鳐是鲨的近亲，体呈扁平状，内骨骼均由软骨所组成。鳃孔开口于腹面，由于常常平贴于海底，容易被泥沙堵塞而造成呼吸困难。为了解决这一问题，它们在眼后方演化出一对圆形的喷水孔，可使水流由此孔流入鳃孔，完成呼吸的全过程。有的种类还能跃出水面，在海面上空作短距离的滑翔。约438种，我国产79种。

Rays is the close relative of shark, the body is flat, and its internal bones are composed by soft bones. The gills are opened on the abdomen, the gills are easily to be jammed by the mud in the sea because sea purse often stick itself to the sea bottom. In order to solve this problem, it has developed a pair of round water-spraying holes, through which the water can flow into the gill hole to complete the breathing process. Some species can even spring out of the water surface and slide a short distance on the surface of the sea. There are totally 438 species, among which our country has 79 species.

双吻前口蝠鲼 *Manta birostris* (Walbaum)
特征：体盘菱形，宽为长的2倍多，尾细长如鞭。
分布：为暖水性中上层鱼类。食游泳甲壳动物或成群小鱼。黄海北部。濒危。

Features: The body is in diamond shape, the width is twice of the length, the tail is slim like a scourge.
Distribution: It is a kind of fish living in the upper middle layer of the ocean. They eat swimming shellfish or fishes in flocks. It distributes in north Huanghai Sea and is severely endangered.

背棘鳐 *Raja clavata* Linnaeus,1758
分布：东大西洋，从挪威和冰岛到大西洋群岛和地中海。中国为新记录。

Distribution: East Atlantic Ocean, from Norway and Iceland to archipelagos and the Mediterranean. There are news records in China.

中国团扇鳐 *Platyrhina sinensis* (Bloch et Schneider)

及达尖犁头鳐 *Rhynbchobatus djiddensis* (Forskal)
分布：南海、东海

Distribution: Nanhai Sea, Donghai Sea shovelnose, sand shark

圆犁头鳐 *Rhina ancylostoma* Bloch et Schneider
分布：印度洋、大洋洲、菲律宾、朝鲜、中国东海、南海。

Distribution: Indian Ocean, Australia, Philippine, Korea, Donghai Sea and Nanhai Sea of China.

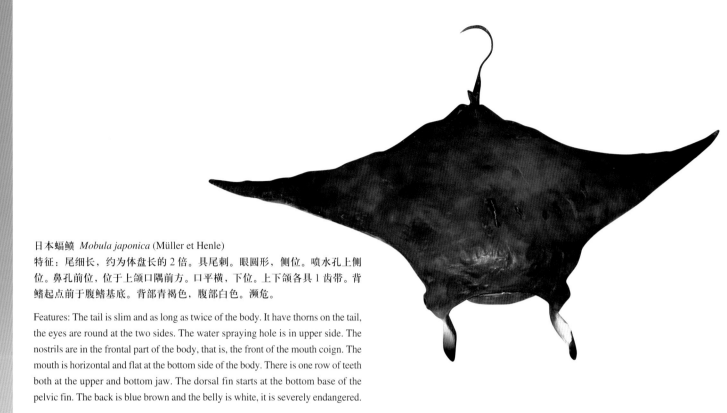

日本蝠鲼 *Mobula japonica* (Müller et Henle)
特征：尾细长，约为体盘长的 2 倍。具尾刺。眼圆形，侧位。喷水孔上侧位。鼻孔前位，位于上颌口隅前方。口平横，下位。上下颌各具 1 齿带。背鳍起点前于腹鳍基底。背部青褐色，腹部白色。濒危。

Features: The tail is slim and as long as twice of the body. It have thorns on the tail, the eyes are round at the two sides. The water spraying hole is in upper side. The nostrils are in the frontal part of the body, that is, the front of the mouth coign. The mouth is horizontal and flat at the bottom side of the body. There is one row of teeth both at the upper and bottom jaw. The dorsal fin starts at the bottom base of the pelvic fin. The back is blue brown and the belly is white, it is severely endangered.

电鳐 Electric Ray

能放电的鱼类中，最有名气的是生活在海洋中的电鳐，它发出的电压，一般大约80伏特，最高的达到目前城市照明用电的水平。电鳐放电主要是为了捕食和自卫。

The most famous fish among all fishes that can discharge electricity is the electric ray in the ocean. Generally it can discharge 80 volt electricity; the highest voltage can reach the illumination level of our city. The electric ray discharge electricity mainly for preying or self-denfending.

丁氏双鳍电鳐 *Narcine timlei* (Bloch et Schneider)

日本单鳍电鳐 *Narke japonica* (Temminck et Schlegel)

鳐类的毒棘 Poisonous Thorns of Ray

为了防御敌害，以守为攻，有些鳐类身上具有特别的装置，此装置包括毒棘与毒腺。被大型魟类毒红棘刺到而致死的例子曾经发生与记载过。

To fight against the enemies and use defense methods to protect themselves, some rays have special organs such as poisonous thorns and glands. Many records state that many people died because they were pricked by the poisonous thorns of rays.

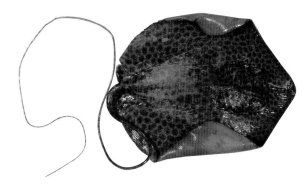

花点魟 *Dasyatis uarnak* (Forskal)

银鲛 (Chimaera)

银鲛是由原始的鲨类演化而来，其躯干侧扁，尾部向后逐渐变细，尾呈鞭状。计有31种，我国有6种。

Chimaera evolved from primitive shark with flat trunk sides, the body becomes slimmer from the tail, which is in scourge shape. There are totally 31 species, our country has 6 species.

黑线银鲛 *Chimaera phantasma* Jordan et Snyser
黑线银鲛为冷温性栖息较深水域中的中小型鱼类。以底栖无脊椎动物为食。
They are medium and small sized fish living in cold deep water area, they prey on ocean bottom invertebrate.

软骨鱼的骨骼 (Bones of Elasmobranch)

虎鲨的骨骼
Bone of horned shark

宽纹虎鲨 *Heterodontum japonicus* Duneril

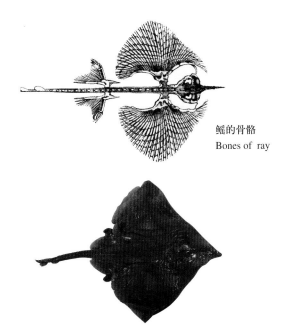

鳐的骨骼
Bones of ray

美鳐 *Raja pulchra* Liu

软骨鱼的卵 (Eggs of Elasmobranch)

现存鱼类中最大的卵是鼠鲨的卵，卵经可达22厘米。刺鳐卵径为9～11厘米。

鲨类一次产出的仔鱼数量是不同的，如角鲨仅7～8尾；吻沟双髻鲨27尾；皱唇鲨约35尾；扁鲨25尾。

The largest eggs among fishes are the eggs of porbeagle, its diameter can be 220mm. The diameters of thorn-back skate is 90~110mm.

Sharks do not produce the same amount of eggs every time, for example, thorn-back skate only produce 7~8 eggs, Scalloped hammerhead can produce 27 eggs, houndshark about 35, and angel shark25.

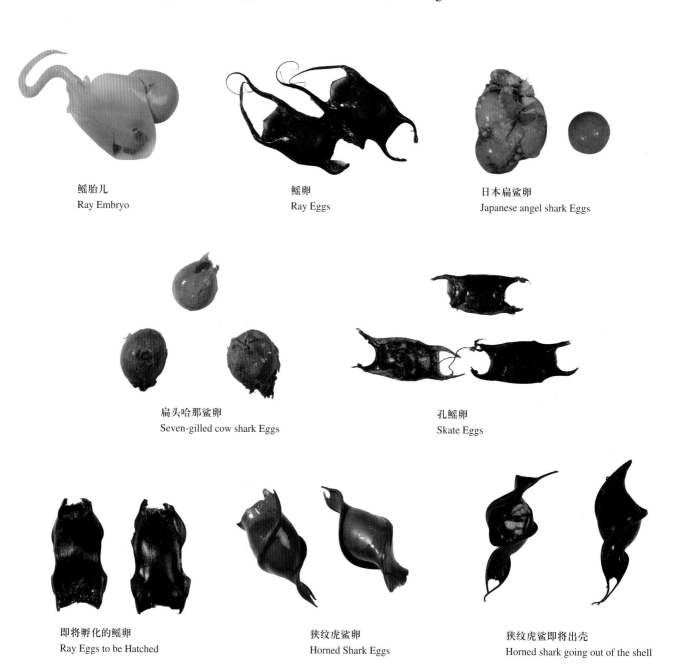

鳐胎儿
Ray Embryo

鳐卵
Ray Eggs

日本扁鲨卵
Japanese angel shark Eggs

扁头哈那鲨卵
Seven-gilled cow shark Eggs

孔鳐卵
Skate Eggs

即将孵化的鳐卵
Ray Eggs to be Hatched

狭纹虎鲨卵
Horned Shark Eggs

狭纹虎鲨即将出壳
Horned shark going out of the shell

海洋硬骨鱼类 (Ocean Teleostean)

硬骨鱼是当今现存鱼类中最大的、也是最繁盛的一个类群，几乎占脊椎动物总数的一半。硬骨鱼类全身的骨骼都是硬骨，口有上下颌，并有成对的偶鳍，具一对鳃孔，大多数有鳔。共有23,600多种，中国产2,950多种。

Teleostean is the largest and most thriving group among all kinds of fishes, it almost accounts for half of all vertebrate. All the bones in the body of teleostean are hard bones, there are upper and bottom jaws, and there are paired fins, they also have a pair of gill holes, most of them have swim bladder. There are totally more than 23,600 species and there are more than 2,950 species in China.

硬骨鱼展厅
The Hall of Teleost

鱼类的内部构造 (Internal Structures of Fishes)

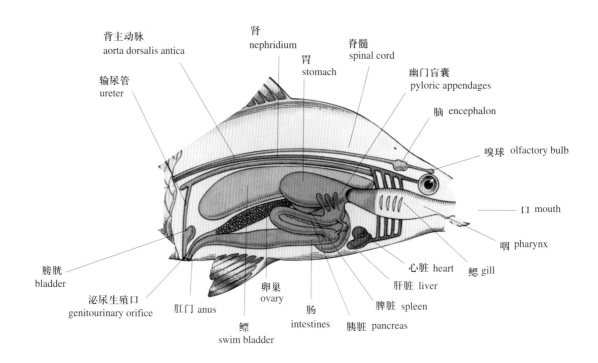

鱼的色彩 (Colors of Fishes)

镊口鱼
Long-snouted coralfish

刺盖鱼
Butterfly fish

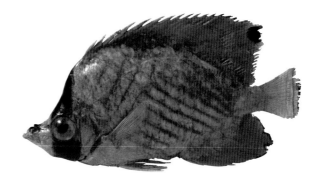

隆头鱼
Lipped fish

双锯鱼
Anemonefish

丝蝴蝶鱼 *Chaetodon auriga* Forskal

鱼的尾和鳍 (The Tails and Fins of Fishes)

适合高速前进的尾鳍
Tail fin suitable for advancing in high speed

不对称的尾鳍
Asymmetrical tail fin

平整的尾鳍
Flat tail fin

展鳍滑翔的胸鳍
Winged gurnard Lepidotrigla

行走在海床上的胸鳍条
Pectoral bar helping walk on sea bed

翼红娘鱼 *Lepidotrigla alata* (Houttuyn)

单棘豹鲂鮄 *Daicocus peterseni* (Nystrom)

珍鲹 *Caranx ignobilis* (Forskal)
背部蓝色，腹部银色；胸鳍腋部深黑色。生活于热带海域，中国产于南海、台湾海峡。此标本是目前本种最大的标本。

It has blue back and silver belly, the pectoral axilla is dark black, it lives in tropical ocean area, and it can be found in Nanhai Sea and Taiwan Straits of China. This specimen is the largest one among this species.

大洋暖水性鱼类 (Fishes Live in Warm Ocean)

东方旗鱼 *Histiophorus orientalis* Temminck et Schlegel

第一背鳍特别高大，帆状。腹鳍较长，仅有一枚鳍棘。吻向前延伸，长而尖，似剑形。头、体背为青蓝色。为热带、亚热带海洋上中层大型凶猛鱼类。

The first dorsal fin is very large in sail shape. The pelvic fin is relatively long; it only has one fin thorn. The long and sharp proboscis in sward shape is protruding forward. The color of its head and back is blue. It is medium sized fierce fish in tropical and semi-tropical ocean.

蓝枪鱼 *Makaira mazara* (Jordan et Snyder)

上颌长为下颌长的1.9～2.4倍，呈枪状伸出。头部和体背铁青色，腹部银白色。为大洋暖水性大型鱼类。游泳迅速，性凶猛。

The length of the upper jaw is 1.9~2.4 times of the bottom jaw in protruding gun shape. The color of head and back is iron blue, the belly is white. It is large fish living in warm ocean. It swims very fast and has a fierce temper.

海洋：生命的摇篮

剑鱼 *Xiphias gladius* Linnaeus.
上颌延长呈剑状突出，以鱼类和乌贼为食。其剑状上颌作用似长尾鲨的尾巴一样，冲入鱼群时，用剑摧残捕获物，而后吞食被打死或被切成碎块的鱼。为大洋暖水性大型鱼类。游泳迅速，性凶猛。中国东海、南海有分布。该标本全长2.10米，1972年采集。

The upper jaw is in sward shape protruding, and it preys on fish and cuttlefish. The functions of its sword shaped upper jaw is the same as the functions of the tails of sharks, when it rushes into flocks of fish, it can use the sword to destroy the victim, and then swallow the dead or smashed fish. It is a large fish living in warm ocean, it can swim in a very fast speed and has a fierce temper. It distributes in the Donghai Sea and Nanhai Sea of China. The total length of the specimen is 2.10m and was taken in the year 1972.

军曹鱼 *Rachycentron canadum*（Linnaeus）
为温带和热带海域的鱼类。中国黄海北部有分布。目前为本种标本之最。

It is a kind of fish living in temperate and tropical oceans. It also distributes in northern Huanghai Sea of China. This is the largest specimen of the species.

鲯鳅 *Coryphaena hippurus* Linnaeus

深海鱼类和珊瑚礁鱼类 (Deep Ocean Fishes and Coral Reef Fishes)

长吻鼻鱼 *Naso unicornis* (Forskal)

帆鳍鱼 *Histiopterus typus* Temminck et Schlegel

新月锦鱼 *Thalassoma lunare* (Linnaeus)

爪哇蓝子鱼 *Siganus javus* (Linnaeus)

云纹蛇鳝 *Echidna nebulosa* (Ahl)

长体鳝 *Thyrsoidea macrurus* (Bleeker)

叉斑钩鳞鲀 *Rhinecanthus aculeatus* (Linnaeus)

横带高鳍鰕虎鱼 *pterogobius zacalles* J. & S.

拟三刺鲀 *Triacanthodes anomalus* (T. & S.)

海鲂 *Zeus japonicus* C. & V.

玻甲鱼 *Centriscus scutus* Linnaeus

烟管鱼 *Fistularia petimba* Lacepede

鲈形鱼类一族 (Perciformes Fishes)

中华马鲛 *Scomberomorus sinensis* (Lacepede)
体长形，侧扁；尾柄细，每侧有3条隆起脊，尾鳍深叉状。多生活在温带海洋，中国东海、南海、黄海有分布。此标本为2004年3月征集，全长2.08米，是目前最大的本种标本。

The body is long and flat in the two sides; the tail is slim with three uprising ridges at each side, the tail fin is deep and in fork shape. It mostly lives in temperate ocean and distributes in the Donghai Sea, Nanhai Sea and Huanghai Sea of China. This specimen was collected in March 2004 with a total length of 2.08 meters; it is the largest specimen of the species in the world.

颊纹双板盾尾鱼 *Callicanthus lituratus* (Bloch & Schneider)

双斑眶棘鲈 *Scolopsis bimaculatus* Ruppell

松鲷 *Lobotes surinamensis* (Bloch)

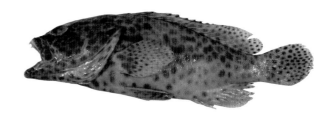
蜂巢石斑鱼 *Epinephelus merra* (Bloch)

舟䲟 *Naucrates doctor* (Linnaeus)

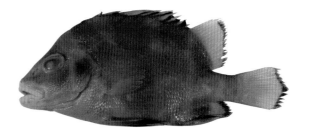
斜带髭鲷 *Hapalogenys nitens* Richardson

细鳞蜊 *Therapon jarbua* (Forskal)

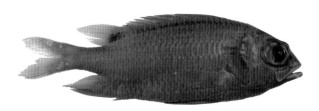
红䱵 *Adioryx tiere* (C. & V.)

条石鲷 *Oplegnathus fasciatus* (Temminck et schlegel)

千年笛鲷 *Lutjanus sebae* (C.&V.)

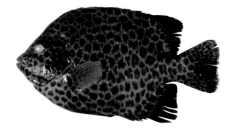
斑石鲷 *Oplegnathus punctatus* (Temminck et Schlegel)

黑尾鹰䲢 *Goniistius guadricoris* (Gunther)

会钓鱼的鱼 (Fish having the Skills of Fishing)

会钓鱼的鱼又称"渔夫鱼",他们的钓竿是第1背鳍棘形的,又称吻触手,其端部有1附属物,又称"拟饵",靠此来诱获那些馋嘴的小鱼到嘴边而吞食之。

The fishing fish is also called "Fisher Fish", their fishing rod is the thorn of the first dorsal fin, it is also called snout antenna with attachment at its end. The attachment is also called "simulated bait", which is used to seduce the piggish small fish.

美国鮟鱇

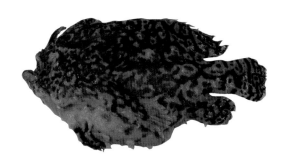
蛙状裸躄鱼 *Histrio ranina* (Tilesius)

棘茄鱼 *Halieutaea stellata* (Vahl)

没有尾鳍的鱼 (Fish without Tail Fin)

翻车鲀 *Mola mola* (Linnaeus)

体短而高，侧扁，椭圆形。口小，前位。上下颌各愈合成1个喙状齿，无中央缝。皮肤粗糙，具粒状突起。背鳍和臀鳍高大，似尖刀状。无尾柄和尾鳍。"舵鳍"边缘波曲，无矛状突起。体背面灰褐色，两侧银白色，各鳍灰褐色。广布于世界各温带及热带海洋。怀卵量约3亿粒，为鱼类之冠。此标本全长2.5米，目前为本种标本之最。

The body is short and high and flat in two sides in eclipse shape. The mouth is small and in the front of the body. Both the upper and bottom jaw develops into a beak-shaped tooth without central slot. The skin is coarse with grain tuber. The dorsal fin and anal fin is very high like sharp knife. It does not have tail stem and tail fin. The edge of the "helm fin" is curved without lance shaped tuber. The body surface is gray and the two sides are silver, each fin is grey brown. It widely distributes in the temperate and tropical ocean. The number of eggs it can impregnate is 300 million, which is the top one among all fishes. This specimen has a length of 2.5 meters, and it is the largest one in the species.

矛尾翻车鲀 *Masturus lanceolatus* (Lienard)

体短卵圆形，侧扁而高。皮肤粗糙，内无骨板。口小，前位。背鳍和臀鳍相似，高而窄，呈尖刀状，鳍条后延，在体后端相连，形成舵鳍，其中央的鳍条延长，呈矛状突起。为热带海洋中广泛分布的漂游性鱼类。

The body is short and round, the two sides are tall and flat. The skin is very coarse without internal bone board. Its mouth is small and in the frontal part of the body. The dorsal fin and anal fin are similar, which is high and narrow in sharp sword shape, the fin strip is retrusive and connected with the rear part of the body, forming steering fin, among which the central fin strip is prolonged like protruding lance. It is an excursing fish widely distributed in tropical ocean.

变态的比目鱼 (Metamorphic Flounder)

变态的鱼种类很多，最典型的是比目鱼。鱼类的变态主要是指体态及体色从幼鱼到成鱼变化较大较明显。如比目鱼，小的时候体对称，随着生长发育，眼逐渐向一侧变移。

There are many kinds of metamorphic fishes; the typical one is the flounder. The abnormalities of fishes mainly refer to the obvious change of posture and body color from parr to fully grown fish such as flounder, it has symmetrical body when it was young, with the development the eyes move to one side of the body.

石鲽 *Kareius bicoloratus* (Basilewsky)

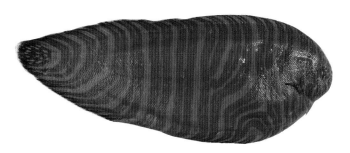

带纹条鳎 *Zebrias zebra* (Bloch)

粒鲽 *Clidoderma asperrima* (Temminck et Schlegel)

高眼鲽 *Cleisthenes herzensteini* (Schmidt)

这是一条异常的个体标本，两眼的位置与正常的高眼鲽相反，又称逆位高眼鲽。

This is an abnormal individual specimen, the positions of the two eyes are opposite to normal position, so it is also called contradictory position pointhead flounder.

其他特殊鱼类 (Other Special Fishes)

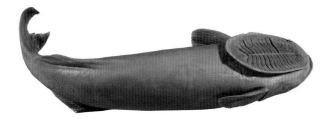

免费旅行者——短鲫 *Remora remoa* (L.)

有毒的鱼——蓑鲉 *Pterois*

条东方鲀 *Takifugu xanthopterus* (T. & S.)

宽尾鳞鲀 *Abalistes stellatus* (Lacepede)

棘背角箱鲀 *Lactoria diaphanus* (Bloch et Schneider)

密斑刺鲀 *Diodon hystrix* Linnaeus

斑鱵 *Hemiramphus far* (Forskal)

凤鲚 *Coilia mystus* (Linnaeus)

硬骨鱼类的繁殖 (The Reproduction of Teleostean)

方氏云鳚的护卵行为
The egg protection behavior of Fang blenny

鳗鲡的发育
The development of Japanese eel

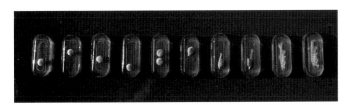

银鲑的发育
The development of coho salmon

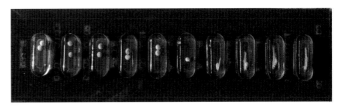

虹鳟的发育
The development of steelhead rainbow trout

可药用的硬骨鱼类 (Teleostean that can be Used in Medicine)

日本七鳃鳗 *Lampetra japonica* (Martens)

尖海龙 *Syngnathus acus* Linnaeus

刁海龙 *Solegnathus hardwicki* (Gray)

大青弹涂鱼 *Boleophthalmus pectinirostris* (Linnaeus)

模式标本及辽宁新记录 (Type Specimens and Newly Recorded Specimens in Liaoning Province)

异形石鲽 *Kareius heteromorpha* sp.nov. 模式标本

分类：鲽形目，鲽科，石鲽属

特征：体侧扁，长椭圆形。头背缘（眼上方）有明显凹刻。两眼位头右侧。口小，右上颌长不及头长三分之一。两颌齿钝锥状，一行。无正常鳞，体右侧沿侧线及上下常有纵行粗骨板，沿背、臀鳍基无一纵行大骨质突起。胸鳍上方侧线半圆凹弧状；右侧无颞上枝。背鳍70，起点基部向前突出，游离状。臀鳍51。肛门位腹缘略偏左侧，生殖突位在肛门后缘稍右。椎骨11 + 27。

生态：同石鲽。

分布：中国黄海北部。

Alien Stone flounder. Type exemplar

Classification: pleuronectiformes, pleuronectidae, platichthys

Features: The body side is flat and the whole body is in long ellipse shape. There are clear concave grooves at the back of the head (above the eyes). The two eyes are at the right side of the body. The mouth is small and the length of the upper right jaw is less than 1/3 of the head. There is one row of teeth which are blunt in taper shape. There is no normal squama and there are vertical thick bone boards along the siding of the right side of the body, and there is no bone tuber along the back and anal fin. The siding above the pectoral fin is in semi-concave arc; there is no superior temporalis at the right side. Dorsal fin 70, the base of the beginning part is protruding forward in dissociation. Anal fin is 51. The anal is slightly right in the belly; the procreation protruding is slight right behind the anal. The vertebra is 11+27.

Distribution: Northern Huanghai Sea in China.

沙塘鳢 *Odontobutis obscurus* (Temminck et Schlegel) 副模式标本

Dark sleeper Subsidiary Type Specimen

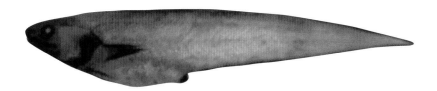

黑潮新鼬鳚 *Neobythites sivicola* (Jordan et snyder) 辽宁新记录标本
特征：背鳍无圆黑斑。体侧无显明的横带状暗色纹。腹鳍条 2。胸鳍22～28，无游离鳍条。长鳃耙7～20，中央基鳃骨牙一群。主鳃盖骨有一强棘。吻长约等于眼径。两腹鳍均位前鳃盖骨下方，相距很近。吻部及下颌无须。匙骨前下段在前鳃盖骨处左右互连。有辅上颌骨。背鳍长于臀鳍。卵生。雄鱼无外插入器官。
生境：为暖温性底层小型鱼类。
分布：旅顺近海；东海。

Whitespotted brotula Newly recorded specimen in Liaoning Province
Feature: There is no round black spot on the dorsal fin. There are no obvious horizontal stripes in dark color. There are two pelvic strips. Pectoral fin is 22-28. There are no dissociation fin strips. Long gill raker is 7-20, there is a group of central basibranchial teeth. There is a strong thorn on the main operculum. The length of the proboscis approximately equals to the length of the eye diameter. The two pelvic fins are all under the frontal operculum and there is only a short distance between the two pelvic fins. There is no hair under the proboscis and the bottom jaw. The frontal bottom part of the cleithrum is connected to both the left and right at the operculum. There are auxiliary upper jaws. The dorsal fins are longer than anal fins. They are born in eggs. The male fishes do not have outer-inserting organs.
Distribution: Lvshun offing; East sea.

鳗鲇 *Plotosus anguillaris* (Bloch,1794) 辽宁新记录标本
特征：体长形，尾渐细。背鳍及臀鳍与尾鳍相连，无脂鳍。口端位，上下颌牙锥形，小须4对。鳃盖膜不与峡部相连。体光滑无鳞，侧线明显。全体棕色，下部较淡。多栖于海湾内之浅沙底，为福建沿海常见的小型底层鱼类，03年9月在大连黑石礁湾采得一尾，为黄海北部新记录（体长242毫米）。体长一般为200～300毫米。产卵于海湾岩礁石缝隙处，产卵期为4～5月。主要摄食底栖生物，如涟虫类、小型短尾类、小型长尾类、瓣鳃类、腹足类和口足类（虾蛄），还摄食一些介型类。

Barbel eel *Plotosus anguillaris* (Bloch,1794) New specimen record in Liaoning Province.
Features: The body is long and gradually becomes slim to the tail. The dorsal fins and anal fins are connected with tail fins, there is no fat fin. The mouth is in the middle position, the upper and bottom teeth on the jaws are in taper shape, the fish also has 4 pairs of small palpus. The operculum is not connected with the isthmus. The body is smooth and has no squama, the sidings are obvious. The whole body is in brown color and the color of the bottom part is slight. They usually inhabit in the shallow sand in bay areas, and they are common small fish in coastal area of Fujian Province, one was caught at Heishijiao, Dalian, it is the latest record in north Huanghai Sea (body length 242 mm). The body length is usually 200~300mm. They lay eggs at the gap in the reefs of the sea; the laying period is from April to May. It mainly preys on animals living in the bottom of the sea, such as cumacea, small fishes with short tails, lamellibranch, gastropod and stomatopoda(squill), they also prey on some animals of ostracod.

大连平鲉 *Sebastes* sp. 模式标本
分类：鲉科
特征：背鳍XIII－12；臀鳍III－6；胸鳍16～17（下部不分枝鳍条9～10，游离状，游离部分约占1/3～1/7）；腹鳍I－5；侧线鳞46～48，12～14/20～24。体侧黑褐布满桔黄色斑点，几乎每一鳞片中部一个，奇鳍上的黄斑较大。头背部黑褐、腹部及胸腹部浅灰褐色，背鳍边缘黑褐。

Dalian rockfish. Type exemplar
Classification: scorpaenidae
Features: The dorsal fins are XIII-12; Anal Fins are III-6; Pectoral fins are 16~17 (There is no difference in unbranched fin-ray 9~10, and they are dissociated, the dissociated part accounts for about 1/3~1/7); pelvic fins I~5; siding squama 46~48, 12~14/20~24. The side of the body is black brown with orange spots, almost one spot for each squama in the middle, the spot is relatively large on the median fin. The head is black brown, the belly and chest is light grey brown, the edge of the dorsal fin is black brown.

大于头长。舌宽阔，游离。体鳗形，无鳞，皮肤光滑；体背侧黑褐色，腹部略淡。齿尖锐细小，口闭时不外露。为近岸底层鱼类。黄海有分布，不常见。可食用。

Japanese conger Newly recorded specimen in Liaoning Province
Features: The sensing hole on the siding is not white; the dorsal fins begin on the upper rear part of the pectoral fin, there is no obvious black spot behind the pectoral fins. It has tail fin which is connected with the back and anal fins. The tail is longer than the total length of the head and trunk. The distance between the anal and gill hole is larger than the head. Its tongue is broad and dissociated. The body is in eel shape and there is no squama, the skin is smooth; the back of the body is black brown, the belly has a lighter color. The tooth tip is sharp and tiny and does not expose when the fish shut the mouth. It lives in the bottom of the offing. It can be found in Huanghai Sea but the chance is very slim. It is edible.

日本康吉鳗 *Conger japonicus* Bleeker 辽宁新记录标本
特征：侧线感觉孔不呈白色；背鳍始于胸鳍后上方，胸鳍后部无显著黑斑。具尾鳍、且与背、臀鳍连续。尾长大于头与躯干合长。肛门至鳃孔的距离

珍稀的鱼 (Rare Fishes)

中华鲟

体具5列骨板，侧骨板40个左右。背部灰黑色，腹部白色，各鳍灰黑色。为中国特有的半洄游性珍贵鱼类。原分布长江口一带。

Chinese Sturgeon
The body has five rows of bone boards and about 40 side bone boards. The back is in grey black, the belly is white and the fins are grey black. It is a rare fish with semi-swimming back habit that particularly lives in China. It originally distributes in the Yangtze River estuary.

中华鲟 *A cipenser sinensis* Gray
保护等级：CITES附录Ⅱ；中国：Ⅰ

Protection Class: CITES Appendix Ⅱ；China: Ⅰ

斑月鱼 Opah

斑月鱼体椭圆而侧扁，红色到紫色，有许多白斑点。体被小圆鳞。胸鳍及腹鳍发达，背鳍前端镰刀状。为上层大洋性鱼类，主要以乌贼、章鱼及甲壳类等为食。分布于太平洋和大西洋。此标本全长1.17米，是目前最大的本种标本。

The body is ellipse and round in two sides, the color is from red to purple with many white dots. The body is covered with small round squama. The pectoral fins and pelvic fins are fully developed; the frontal part of the dorsal fin is in sickle shape. It is upper layer ocean fish, and mainly preys on cuttlefish, octopus and shellfish. It distributes in Pacific Ocean and Atlantic Ocean. The Specimen is 1.17 meters long and it is the largest specimen of the species.

斑月鱼 *Lampris guttatus* (Brunnich)

南极鱼 *Notothenia squamifrons* Gunther

文昌鱼 *Branchiastoma*
保护等级：中国：Ⅱ

Protection Class: China: Ⅱ

松江鲈 *Trachidermus fasciatus* Heckel
保护等级：中国：Ⅱ

Protection Class: China: Ⅱ

海洋无脊椎动物 (Ocean Invertebrates)

　　海洋无脊椎动物主要包括原生动物、海绵动物、腔肠动物、扁形动物、线形动物、环节动物、软体动物、节肢动物、棘皮动物等。

Ocean invertebrates mainly include protozoa, poriferan, coelenterate, platyhelminth, Nemathelminthes, annelid, mollusk, arthropod and echinoderm, etc.

海洋无脊椎动物展厅
The Hall of Marine Invertebrates

海洋无脊椎动物展厅
The Hall of Marine Invertebrates

海洋无脊椎动物展厅
The Hall of Marine Invertebrates

多孔动物——海绵 (Poriferan-Sponges)

海绵，因其身体柔软，大都生活在海洋里而得名。又因其身体表面有无数小孔，也称为多孔动物。海绵动物是最原始、低等的多细胞动物，约有1万多种。体形呈管状、球状、树枝状、瓶状等。

Sponge gets its name because it has a soft body and lives in the ocean. It is also called poriferan because there are many tiny holes in its body. Sponge is the most primitive and low cellulous animal, there are totally more than 10,000 species. Its body can be cannular, bulbiform, in branch shape and bottle shape, etc.

偕老同穴海绵　生活在深海里，体形呈长圆笼状，体内常常居住着一对雌雄小虾，这种小虾自幼就进入海绵体腔，长大后被海绵骨针阻隔在内不能出来，只好在海绵体内白头偕老。

It lives in deep ocean with long round body, usually there will be a male and a female shrimp living in its body, the shrimps enter the celom of the sponge since they are very young, after they grow up they cannot come out of the sponge because they are blocked inside by the spicules of the sponge, and they can only live to death inside the sponge.

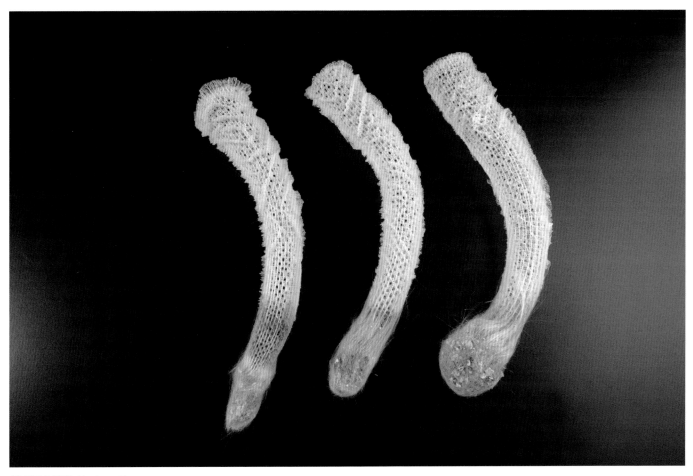

偕老同穴海绵 *Eupleectella* sp.

美丽的珊瑚 (Beautiful Corals)

　　珊瑚属腔肠动物，是由珊瑚虫组成的一簇簇不定型的群体结构。珊瑚虫能分泌角质或石灰质的外骨骼，在珊瑚骨骼上有许多小孔，每个小孔内都居住着一个珊瑚虫。

　　珊瑚分布于热带和亚热带海洋中，按生活环境分为造礁珊瑚和非造礁珊瑚。造礁珊瑚的石灰质骨骼形成了千姿百态的珊瑚礁。

Corals belong to coelenterate, they are a colonial structure without fixed shape formed by actinozoan which can excrete calcareous exoskeleton, there are many tiny holes in the bones of the coral, and each of the holes is a residence of one actinozoan.

Coral lives in tropical and semi-tropical oceans, according to their living environment they can be divided into reef-building corals and non-reef building corals. The calcareous bones of reef building corals compose various coral reefs.

牡丹珊瑚 *Pavona* sp.
保护等级：CITES：附录 II

Protection Class: CITES: Appendix II

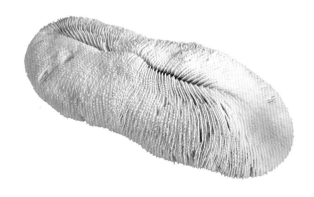

刺石芝珊瑚 *Fungia echinata*
保护等级：CITES：附录 II

Protection Class: CITES: Appendix II

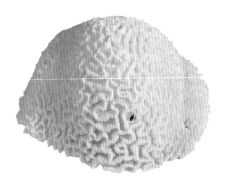

脑珊瑚 *Platygyra* sp.
保护等级：CITES：附录 II

Protection Class: CITES: Appendix II

笙珊瑚 *Tubipora musica*
保护等级：CITES：附录 II

Protection Class: CITES: Appendix II

石芝珊瑚 *Fungia fungites*
保护等级：CITES：附录 II

Protection Class: CITES: Appendix II

多姿多彩的海洋贝类 (Rich and Colorful Seashells)

贝类属软体动物，主要包括单壳类、双壳类、多板类、掘足类、头足类等。其共同特征是身体柔软不分节，由头、足、内脏囊、外套膜和贝壳组成。

Seashells belong to mollusk, they mainly include univalve, bivalve, chitons, tusk shells and cephalopod, etc. Their common features are that their bodies are soft and there are no segments in their bodies which are composed of head, feet, visceral bursa and the shells.

千姿百态的螺类 (Multifarious Spiral Shells)

螺类属于软体动物腹足纲，通常有一个螺旋形的贝壳，足部发达，位于身体的腹面，所以在分类上统称为"腹足类"。

Spiral shells have shells in spiral shape, their feet are fully developed at the side of the body, therefore they are classified as "gastropod".

卵梭螺
Ovula ovum

皱纹盘鲍
Haliotis discus hannai

夜光蝾螺
Turbo marmoratus

驼背扭螺
Distorsio anus

带蝾螺
Turbo petholatus

鹬嘴骨螺
Murex haustellum

金塔玉黍螺
Tectarius coronatus

太阳衣笠螺
Xenophora solaris

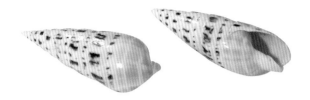

罗纹笋螺
Terebra maculata

红口榧螺
Oliva miniacea

唐冠螺 *Cassis cornuta*
保护等级：中国：Ⅱ

Protection Class: China: Class Ⅱ

南非蝾螺
Turbo sarmaticus

万宝螺
Cypraecassis rufa

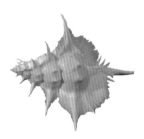

岩石芭蕉螺
Chicoreus alabaster

栉棘骨螺
Mures pecten

缀壳螺
Xenophora pallida

分层笋螺
Terebra dimidiata

大马蹄螺
Triochus niloticus

西兰犬齿螺
Vasum ceramicum

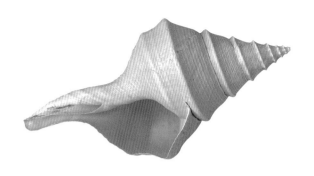

澳大利亚香螺
Syrinx aruanus

法螺
Charonia tritonis

奇形怪状的凤螺

凤螺外唇前端有呈"U"字形的缺刻，可以使左眼伸出壳外。凤螺约有100多种，大多生活在热带海洋中。

Japanese ivory shell that is Grotesque in Shape

There is a "U" shape concave outer frontal edge which can make the left eye pull out of the shell. There are more than 100 species of Japanese ivory shell, most of which live in tropical ocean.

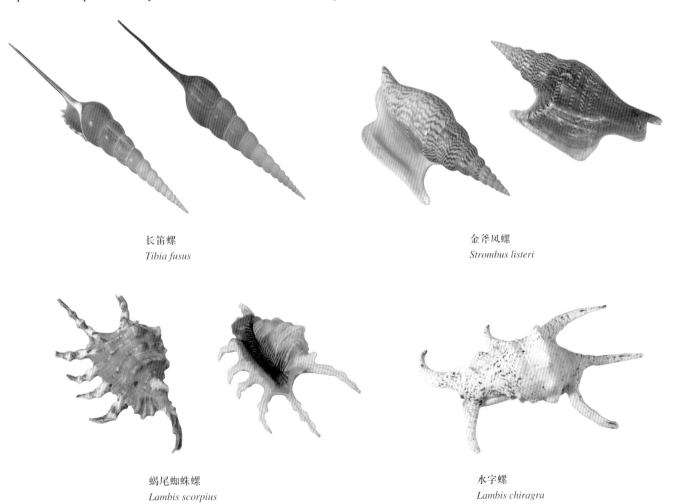

长笛螺
Tibia fusus

金斧凤螺
Strombus listeri

蝎尾蜘蛛螺
Lambis scorpius

水字螺
Lambis chiragra

色彩鲜艳的芋螺 Colorful Taro Shell

芋螺贝壳呈倒锥形，形状有些像人们吃的芋头或鸡的心脏，所以又叫鸡心螺。中国约有70多种。

Taro Shell is in converse taper, the shape is similar to taro that we eat and the chicken heart, therefore it is also called chicken-heart shell. There are totally about over 70 species in China.

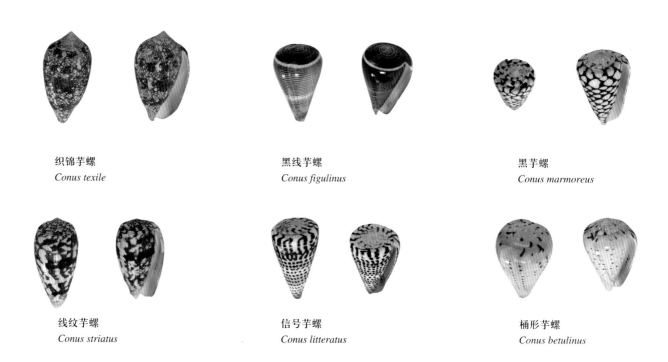

织锦芋螺 *Conus texile*

黑线芋螺 *Conus figulinus*

黑芋螺 *Conus marmoreus*

线纹芋螺 *Conus striatus*

信号芋螺 *Conus litteratus*

桶形芋螺 *Conus betulinus*

绚丽的宝贝 Flowery Cowrie

宝贝是最为艳丽的贝类，其贝壳近于卵圆形，分布于热带和亚热带海域。

The cowrie is the most flowery shell, it's almost in egg shape and it distributes in tropical and semi-tropical maritime space.

玛瑙拟枣贝 *Erronea onyx*

蛇目瓹贝 *Talparia argus*

酒桶宝贝 *Cypraea atlpa*

绶贝 *Mauritia mauritiana*

图纹绶贝 *Mauritia mappa*

龟甲贝 *Chelycypraea testudinaria*

宝贝的外套膜从贝壳腹面两侧反转向上把贝壳包被起来，当宝贝活动时，外套膜经常分泌珐琅质涂在壳上，使贝壳变得光滑艳丽。

The outer membrane covers the shell from the two sides of the belly to the reverse upper side of the cowrie, when it is active, the outer membrane usually excrete enamel on the shell to make it smooth and attractive.

虎斑宝贝生态图
Tiger Spot Cowrie Ecological Picture

虎斑宝贝 *Cypraea tigris*
保护等级：中国：Ⅱ

Protection Class: China: Class Ⅱ

虎斑宝贝又叫黑星宝螺或虎皮贝。因壳表面上有酷似虎豹身上的斑点，故有虎斑宝贝之称。中国的台湾、海南岛、西沙群岛、南沙群岛均有分布。

Tiger spot cowrie is also known as Xingbao Shell or Hupi Shell, there are spots on the shell surface like the spots on tiger and leopard, so it is called tiger spot cowrie. It distributes in Taiwan, Hainan Island, Xisha Archipelago, and Nansha Archipelago in China.

卵黄宝贝
Cypraea vitellus

山猫眼贝
Cypraea lynx

色彩缤纷的双壳贝类 (Bivalve of Various Colors)

双壳贝类具有2枚贝壳，它们的身体左右侧扁，鳃通常呈瓣状，所以也叫"瓣鳃类"。其头部退化，足发达呈斧头状，故又叫"无头类"或"斧足类"。

Bivalves have two shells, their bodies are flat in left and right side, the gill is usually in petal shape, so they are also called lamellibranches. The head is degenerated, the feet is fully developed into ax shape, so they are also called acephala or pelecypoda.

企鹅珍珠贝
Pteria penguin

锯齿牡蛎
Lopha cristagalli

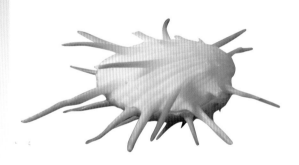

瑞氏海菊蛤
Spondylus wrightianus

砗磲 *Hippopus hippopus*
保护等级：CITES：附录 Ⅱ

Protection Class: CITES: Appendix Ⅱ

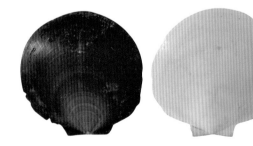

日本日月贝
Amusium sp.

须毛海菊蛤
Spondylus sp.

贝类之王——砗磲 (King of the Shells-Griant clam)

砗磲是有壳贝类中的"巨人"和"寿星"。贝壳长可达1.8米，重量可达200公斤，寿命可达百年。砗磲共有6种，中国西沙群岛均有分布。

Griant clam is the "giant" and "the long-living one" among animals with shells. It can be as long as 1.8m and as heavy as 200 kg, it can live up to a hundred years. There are totally 6 species of Griant clam which can be found in Xisha Archipelago of China.

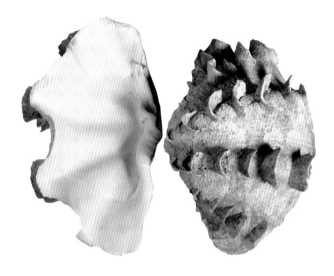

鳞砗磲 *Tridacna squamosa*
保护等级：CITES：附录 II

Protection Class: CITES: Appendix II

砗磲的外壳可作浴盆使用
The outer shell of Griant clam can be used as bath tub

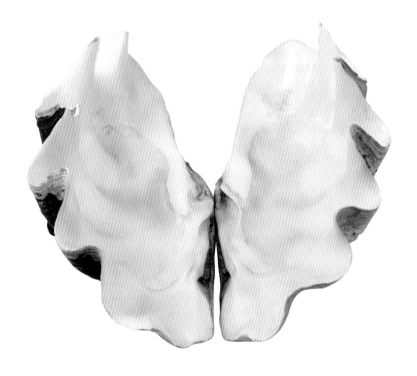

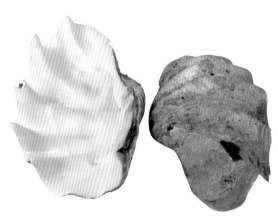

库氏砗磲 *Tridacna cookiana*
保护等级：中国：I；CITES：附录 II

Protection Class: China: I ; CITES: Appendix II

珍贵的鹦鹉螺 (Precious Nautilus)

鹦鹉螺是具有外壳的头足类软体动物,因其贝壳表面具有红褐色的波状条纹,形如美丽的鹦鹉而得名。主要分布在热带和亚热带海域,中国台湾和海南岛有分布。

Nautilus is mollusk of cephalopod, its shell surface is covered with red brown wave shaped stripes, it seems like beautiful parrot, and thus the shell got the name "nautilus". It mainly distributes in the tropical and semi-tropical maritime areas, it can also be found in Taiwan and Hainan Island of China.

鹦鹉螺壳的横切面
Transverse Section of the Shell of Nautilus

鹦鹉螺 *Nautilus pompilius*
保护等级:中国:Ⅰ

Protection Class: China: Ⅰ

身穿"盔甲"的动物——甲壳动物 (Animals with "Corselet"-Shellfish)

甲壳动物是海洋中常见的节肢动物，其身体外面具有几丁质外骨骼。种类很多，体形大小相差悬殊，如蜘蛛蟹两螯展开宽可达4米，水蚤则需借助显微镜才能看清。

Shellfishes widely distribute in the ocean, they have chitin outer bones. There are many species of shellfish with tremendous differences in sizes, for example, the spider crab can be as long as 4 meters when it spreads its two pincers, however the daphnia can be seen only when we use a microscope.

窄琵琶蟹
Lyreidus stenops

红斑斗蟹
Liagore rubromaculata

红星梭子蟹
Portunus sanguinolentus

中华虎头蟹
Orithyia siinica

锦绣龙虾
Panulirus ornatus

绵蟹
Dromia dehaani

蛙形蟹
Ranina ranina

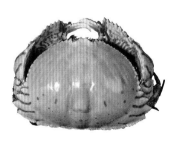

卷折馒头蟹
Calappa lophos

海龟 (Marine Turtles)

生活在海里的龟，个体都比较大，身体呈流线形，四肢呈桨状，适应海洋生活。

Turtles living in ocean have large bodies in stream line shape; the limbs are in paddle shapes which are suitable for marine life.

棱皮龟 *Dermochelys coriacea*
保护等级：中国：Ⅱ；CITES：附录Ⅰ

Protection Class: China: Ⅱ；CITES: Appendix Ⅰ

海龟 *Chelonia mydas*
保护等级：中国：Ⅱ；CITES：附录Ⅰ

Protection Class: China: Ⅱ；CITES: Appendix Ⅰ

蠵龟 *Caretta caretta*
保护等级：中国：Ⅱ；CITES：附录Ⅰ

Protection Class: China: Ⅱ；CITES: Appendix Ⅰ

玳瑁 *Eretmochelys imbricata*
保护等级：中国：Ⅱ；CITES：附录Ⅰ

Protection Class: China: Ⅱ；CITES: Appendix Ⅰ

藻类 (Algae)

美丽一族——红藻 (Beautiful Group-Red Algae)

红藻全世界有4000多种，其中只有1.2%的种类生活在淡水环境中。它们是海洋中最美丽的植物，也是别具特色的经济植物。

Red algas are totally of more than 4,000 species in the world, among which 1.2% live in freshwater. They are the most beautiful plants in the ocean, they are also the most special economic plants.

角叉菜 *Chondrus* sp.
角叉菜含有丰富的卡拉胶，可作为工业上提取卡拉胶的原料；可食；可作为中草药，具有润肠通便、和血消肿、止痛生肌等功效。

It contains rich carrageenan and can be used as raw material in industrial carrageenan extraction; it is edible; it can be used as traditional Chinese herbal medicine with the functions of embellishing intestines and helping defaecation, adjusting blood condition and helping detumescence, and pain killing and helping new muscle grow.

石花菜 *Gelidium* sp.
石花菜在中国、日本和东南亚各地都有悠久的食用史；是中草药，可治疗肾炎、肠炎等病症；是工业上提炼琼胶的好原料。

It has a long edible history in China, Japan and Southeast Asia; it is traditional Chinese herbal medicine that can be used for nephritis and enteritis, etc., it is also good material for industrial agarose extraction.

珊瑚藻 *Corallina officinalis*
珊瑚藻全身充满钙质，是中药"浮石"的原植物，具有良好的驱蛔虫作用。

The body is full of calcium, it is the original plant for traditional Chinese medicine pumice which is very good at belly worm driving.

蜈蚣藻 *Grateloupia filicina*

多管藻 *Polysiphonia urceolata*

金膜藻 *Chrysymenia wrightii*

绒线藻 *Dasya villasa*

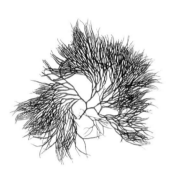
滑枝藻 *Tsengia nakamurae*

海索面 *Nemalion helminthoides* var. *vermiculare*

钩凝菜 *Campylaephora hypnaeoides*

粗枝软骨藻 *Chondria crassicaulis*

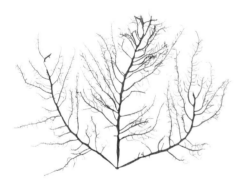
蠕枝藻 *Helminthocladia yendoana*

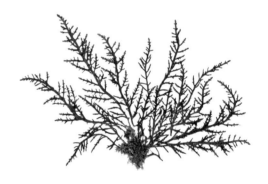
凹顶藻 *Laurencia* sp.

鹿角海萝 *Gloiopeltis tenax*

三叉仙菜 *Ceramium kondoi*

经济植物之家——褐藻 (Economic Plant Family-Brown Algae)

褐藻有1500多种，其中大部分生活在寒带或南、北极的海洋中。它们中经济植物居多，不仅有闻名世界的巨藻、海带和裙带菜，还有诸多鲜为人知而又有经济价值的种类。

Phaeophyta are totally more than 1,500 species, most of them grows in frigid zone or the oceans of south pole or north pole. Many of them are economic plants including world famous Macrocystis pyrifera, Laminaria japonica, Undaria pinnatifida and other species which are seldom known but have much economic value.

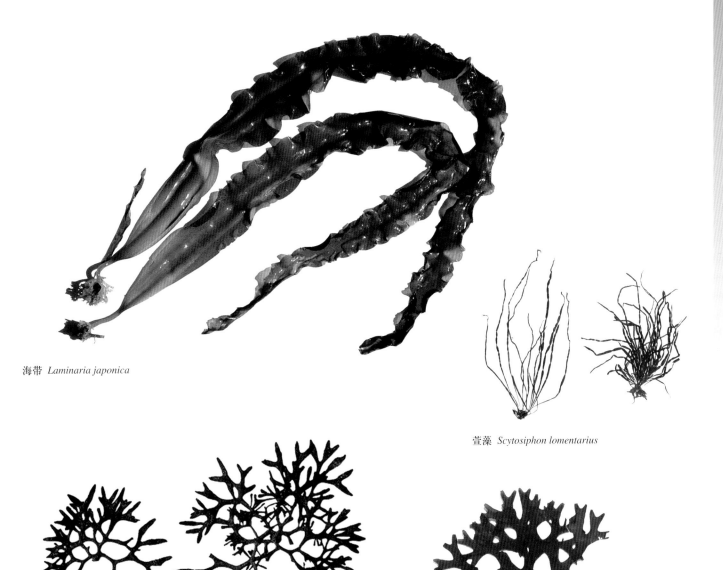

海带 *Laminaria japonica*

萱藻 *Scytosiphon lomentarius*

鹿角菜 *Pelvetia siliquosa*

叶状铁钉菜 *Ishige foliacea*

叉开网翼藻 *Dictyopteris divaricata*

褐舌藻 *Spatoglossum pacificum*

点叶藻 *Punctaria plantaginea*

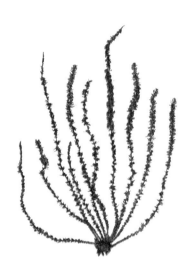
鼠尾藻 *Sargassum thunbergii*

海蒿子 *S. pallidum*

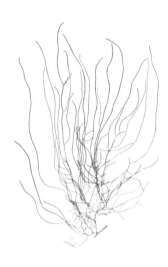
海蕴 *Nemacystus decipiens*

铜藻 *S. horneri*

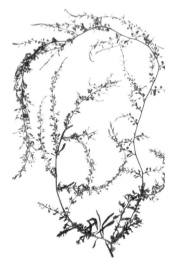
海黍子 *S. kjellmanianun*

羊栖菜 *Hizikia fusiforme*

潮间带的"绿金"——绿藻（Tidal Zone "Green Gold"-Green Algae）

绿藻有8600多种，其中90%生活在淡水中，只有10%生活在海洋的潮间带。

Green algas are totally more than 8,600 species, 80% of them live in freshwater, only 10% live in the tidal zone of the ocean.

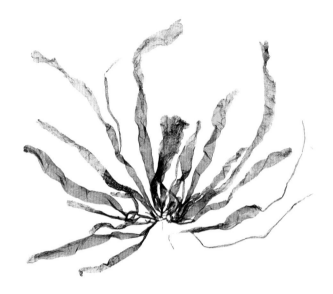

浒苔 *Enteromorpha* sp.
自古就是群众喜欢的海菜，不仅鲜菜可以食用，而且以干菜为原料加工的苔菜粉是良好的食品添加剂。

Many people love to eat it since ancient time, the fresh plant can be eaten, and in addition, the alga powder is also very good food additive.

石莼 *Ulva* sp.
石莼是潮间带最常见、产量最高的藻类，可鲜食，也可晒干作食品添加剂。具有多种保健作用。它也是良好的家禽饲料。

Ulva sp. is the most common and productive alga in tidal zone, it can be eaten fresh or food additive after dried. It has many functions in health keeping. It is also very good domestic bird food.

刺松藻 *Codium fragile*
刺松藻是大型单细胞海藻，整个藻体仅仅由一个细胞组成，只是细胞内不是一个细胞核，而是有无数个细胞核。

Codium fragile is large one-celled alga, the whole body is composed by one cell, but there are numerous karyons inside the cell.

哺乳动物 (Mammals)

哺乳动物是动物界进化地位最高的自然类群，其特点是身披毛发、体温恒定、泌乳育婴。

Mammals are the most advanced evolution group in the animal kingdom, their features are that they are covered with hairs, and they have invariable temperatures and can breed babies by their milk.

现存哺乳动物间可能的相互关系
Possible Relationships between Present Mammals

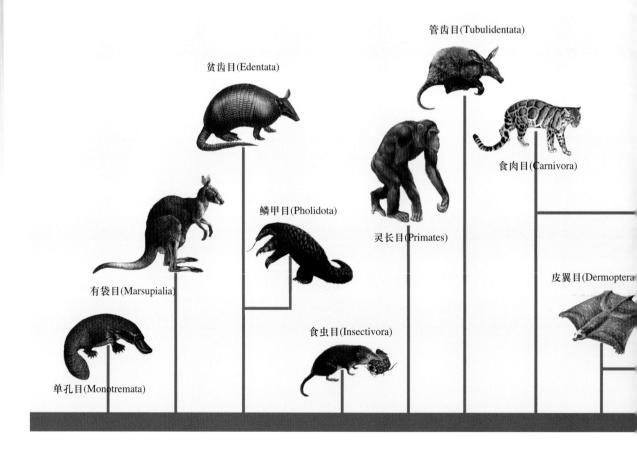

哺乳动物演化时间表
Mammals Evolution Timetable

中生代 Mesozoic			新生代（第三纪）Cenozoic (Tertiary)					第四纪 Quaternary	
230	195	141	66	55	37.5	24	5	1.7	0.01
三叠纪 第一批哺乳动物出现，爬行动物继续增多 The Triassic Mammals appeared. The reptiles increased unceasingly.	侏罗纪 第一批鸟类出现，恐龙全盛时期 The Jurassic The first birds appeared. The epoch is the dinosaur's florescence.	白垩纪 哺乳动物及鸟类增多，恐龙日见稀少直至灭绝 The Cretaceous Mammals and birds increased. Dinosaur decreased gradually until to die out.	古新世 哺乳动物急速分化，但与现代哺乳动物仍然有很大差别 The Paleocene The mammals diversified rapidly, but still differed from present mammals.	始新世 第一批灵长类动物及蝙蝠出现，早期马出现 The Eocene The first primates and bats appeared. Early horses appeared.	渐新世 第一批乳齿象及犀牛的近亲出现 The Oligocene The mastodon appeared. The close relatives of rhinoceros appeared.	中新世 猿类出现，具有现代形态的草食哺乳动物逐渐繁盛 The Miocene Apes appeared. The herbivore with modern shape flourished gradually.	上新世 人类出现 The Pliocene Human beings appeared.	更新世 冰河时期数度来临，适应冰河的动物增加 The Pleistocene The glacial period came several times, letting animals accustomed to glaciers increased.	全新世 现代哺乳动物及人类在各大陆上急速增加 The Holocene Modern mammals and human beings increased rapidly on continents.

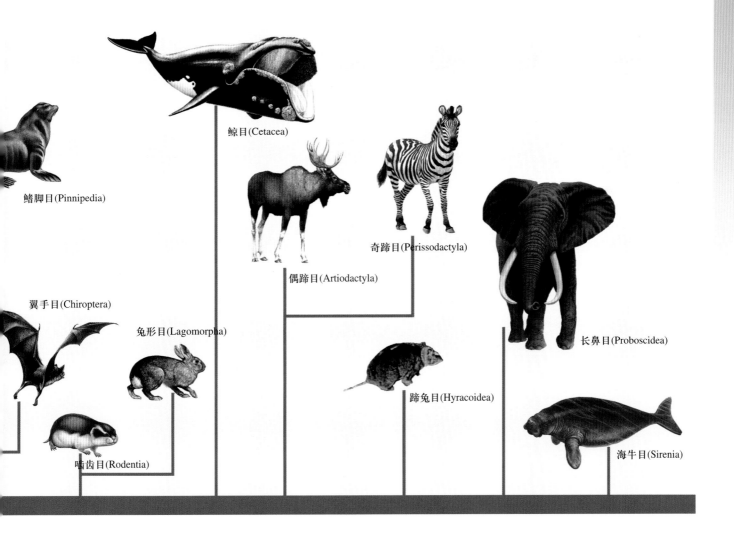

大连自然博物馆

物种多样性展厅
The Hall of Biodiversity

东北森林动物展厅
Northeast forest animal hall

哺乳动物的繁殖 (Reproducing of Mammals)

下蛋的哺乳动物——单孔类 (Mammals Laying Eggs-Monotremata)

由于卵生哺乳动物的粪便、尿、卵都由一个排泄孔出来，故称单孔类。

Mammals which breed in eggs, excrete dejection, urination and eggs in one excretive hole, therefore they are called Monotremata.

鸭嘴兽 Duckbill

鸭嘴兽的嘴似鸭子，足有蹼，体表披毛，变温。居住在河川、湖泊堤边，挖洞筑巢产卵。

Mouth similar to duck; Webs on the feet; Body is covered with hair and body temperature changes. Resident in banks of river and lake; Lays eggs in holes dug by itself.

鸭嘴兽 *Ornithorhynchidae anatinus*

鸭嘴兽的分布
Distribution of Duckbill

针鼹 Echidna

针鼹体表披针毛，卵产在育儿袋中。没有牙齿，用长舌猎食昆虫。生活在干燥草原及森林。

Body is covered with sharp quills and lays eggs in the pouch. It does not have teeth and can only use the long tongue for hunting. It mainly lives in dry grassland and forest.

针鼹 *Tachyglossus aculeatus*
保护等级：CITES 附录 I

Protection Class: CITES Appendix I

针鼹分布
Distribution of Echidna

有育儿袋的哺乳动物——有袋类 (Mammals with Pouches-Marsupialia)

有袋类动物是胎生，几乎没有胎盘，雌兽的腹部有育儿袋、乳头，婴儿在育儿袋中继续发育。

They reproduce by viviparity and almost do not have placenta, there are pouches and nipples at the abdomen of the female animal, the babies continue to develop in the pouch.

大灰袋鼠 *Macropus giganteus*

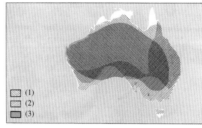

1. 毛袋鼠 Rock Kangaroo.
2. 灰袋鼠 Grey Kangaroo.
3. 红袋鼠 Red Kangaroo

袋鼠的分布
Distribution of Macropus

有胎盘哺乳动物 (Mammals with Placenta)

胎儿在母体子宫内发育，叫胎盘哺乳动物。

The embryo develops in the uterus of the mother, they are called placenta mammals.

印度黑羚的分布
Distribution of Antilope cervicapra

印度黑羚 *Antilope cervicapra*
保护等级：CITES: 附录Ⅲ

Protection Class: CITES: Appendix Ⅲ

哺乳动物的自卫 (Self-defense of Mammals)

身披铠甲的哺乳动物 (Mammals with Loricae)

九绊犰狳 Nine-banded Armadillo

九绊犰狳全身覆盖鳞片。当遭到攻击时，将背拱起，保护柔软的腹部，或挖洞躲藏。

Body is covered with squama. When attacked, Armadillo raises the back to protect soft belly, or digs holes to hide in.

穿山甲 Chinese Pangolin

穿山甲的头、身体、尾巴和腿都覆盖着大而平的角质鳞片。当遇到麻烦时，就将身体团成一团，然后用宽宽的尾巴盘住脑袋。

Head, back, tail and legs are covered with large and flat squama. When threatened contracts into a mass, uses the broad tail to cover the head.

九绊犰狳 *Dasypus novemcinctus*

穿山甲 *Manis pentadactyla*
保护等级：中国：Ⅱ；CITES：附录Ⅱ

Protection Class: China: Ⅱ; CITES: Appendix Ⅱ

身披长刺的哺乳动物 (Mammals with Quills)

豪猪 Short-tailed Porcupine

豪猪身披长矛。当被攻击时，矛刺竖起，继而转身以背刺相向，倒退冲刺，将矛刺戳进攻击者的皮肤。

Body are covered with long sharp quills. When threatened erect quills, run backwards into attackers; and quills are often left embedded in the predator's flesh.

刺猬 Hedgehog

刺猬除腹部没有针刺外，全身披有针刺。当它蜷起身子时，缩紧全身肌肉，形成一个圆球，外面是张开的硬刺。

Body except abdomen cover with hard spines. In curling up crimples the muscles to form a ball with hard spines on the surface.

豪猪 *Hystrix brachyura*

刺猬 *Erinaceus amurensis*

头上长角的哺乳动物 (Mammals with Horns or Antler)

角是有蹄类的防卫利器,常见的有洞角及实角。洞角为头骨的骨角外面套以由表皮角质化形成的角质鞘构成;洞角不分叉,终生不更换。实角是由真皮骨化后,穿出皮肤而成的骨质角;实角分叉,每年脱换一次。

Horns or Antlers are defense weapons for ungulate. Horns are formed by the cutin sheath in the keratinization process of the cuticle and the outer shell of the skull; hollow, no pronged and no shed in the lifetime. Antlers are formed after the ossification of the corium, pronged and shed once a year.

1. 狍子 *Capreolus capreolus* 角分三叉,角干多结节,为实角。
2. 叉角羚 *Antilocapra americana* 角为套状外壳。
3. 汤氏瞪羚 *Gazella thomsonii* 角为洞角。
4. 白犀 *Ceratotherium simum* 角是由毛聚成的。
5. 长颈鹿 *Ciraffa camelopardalis* 角终生包被皮毛。

1. Roe Deer   Antlers with many knurs develop three tines.
2. Pronghorn   Having hollow, pronged, shed horns.
3. Thompson gazelle   Having hollow horns.
4. White Rhinoceros   Horns are composed of hairs.
5. Giraffe   "Horns" are not true horns but knobs covered with skin and hair during the lifetime.

哺乳动物的尾巴 (Tails of Mammals)

河狸 Eurasian beaver

河狸尾宽大而扁平,覆盖角质鳞片,在水中用来做控制方向的舵;遭遇敌害时,通过上下击水警告同伴。

Flat and broad tail covered with cutin squama, can be used as steering to control direction in water. When attacked, tail slaps water for warning.

北极狐 Arctic Fox

北极狐大尾巴就像一条围巾,可用来围住自己的身体。尾尖毛束的颜色可作为与同伴联系的视觉信号。

Large tail is similar to a scarf, can cover the whole body. The color of the tip can serve as vision signal to associates.

河狸 *Castor fiber*
保护等级:中国:I
Protection Class: China: I

北极狐 *Alopex lagopus*

斑马 Common Zebra

斑马尾巴由数百根又长又粗的毛组成,能用来赶走缠身的蚊蝇和小虫。

Tail composed of several hundred long and thick hairs, can drive away bothering mosquitos, flies and small insects.

松鼠 Common Squirrel

松鼠尾巴很大,当它从树上跳下时,像降落伞一样,使松鼠安全地落下来,落地时,松软的尾巴又可起到海绵垫的作用。

In jumping off the tree, the very large tail serves as a parachute to ensure landing safe, and the soft tail acts as sponge pad.

斑马 *Equus burchelli*

松鼠 *Sciurus vulgaris*

赤颈袋鼠 Red-necked Wallaby

赤颈袋鼠尾长而粗壮,休息时可支撑身体,奔跑时可做平衡杆。

Thick and strong tail can support the body in resting and serve as balancing pole in running.

赤颈袋鼠 *Macropus rufogriseus*

哺乳动物的牙齿 (Teeth of Mammals)

哺乳动物因食物及进食方式的不同，发展出具备切割、嵌夹、穿刺、研磨等不同功能和形状的牙齿。门齿有切割食物的功能，犬齿有撕裂的功能，臼齿具有咬、切、压、研磨等多种功能。

Because of the difference in food choice and food taking, the mammals have developed teeth in different shapes and with different functions for cutting, embedding, piercing and rubbing. The fore-tooth can cut food, canine has the function of tearing, the cheek tooth can bite, cut, press and rub.

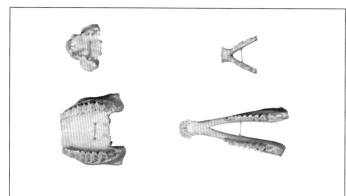

草食动物和肉食动物的牙齿的比较
Teeth Comparison between Plant-eating Animals and Flesh-eating Animals

几种动物的头骨与牙齿比较
Skull and Teeth Comparison of Several Animals

小灵猫 *Viverricula indica* 中国：Ⅱ
China: Ⅱ

红颊獴 *Herpestes javaniaus*

狼獾 *Gulo gulo*

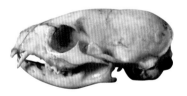

缺齿鼹 *Mogera robusta*

黄鼬 *Mustela sibirica*

貉 *Nyctereutes procyonoides*

水獭 *Lutra lutra*

紫貂 *Martes zibellina*

赤狐 *Vulpes vulpes*

哺乳动物的飞行 (The Flying of Mammals)

哺乳动物中只有蝙蝠才真正具有飞行本领，鼯鼠等只能利用身上的皮膜在空中滑翔。
Among all mammals, only the bat has the skill of flying. Flying squirrel can only slide in the air utilizing the involucra on its body.

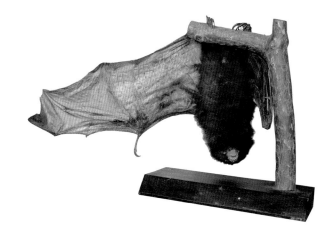

狐蝠 *Pteropus dasymalus*
保护等级：CITES：附录 II

Protection Class: CITES: Appendix II

高度进化的灵长类动物 (Advanced Primates)

灵长类动物包括猴、猿、猩猩以及人类。它们是哺乳动物中进化程度最高的，具有比较发达的大脑，复杂的行为，灵敏的听觉、触觉和视觉。

Primates include monkey, ape, orangutan and human beings. They are the most advanced mammals and they have fully developed brain, complex behavior, acute hearing, touch and vision.

金丝猴 Golden monkey

金丝猴是中国特有的猴类。雄猴背部外被细而密的金色长毛，达30厘米，故名金丝猴。因其头圆，耳短，鼻孔上仰，又称仰鼻猴；又因其面部天蓝色，又称蓝面猴。

It is a monkey particularly living in China. The back of male is covered with golden hair as long as 30 cm, therefore it is named Golden monkey. It is also called Snub-nosed Monkey because the head is round with short ears and upturning nostril; It can also be called Blue-faced Monkey because of its blue face.

金丝猴 *Pygathrix roxellana*
保护等级：中国：I；CITES 附录 I；IUCN：VU

Protection Class: China: I ; CITES Appendix: I ; IUCN: VU

蜂猴 *Nycticebus coucang*
保护等级：中国：Ⅰ；CITES：附录Ⅱ

Protection Class: China: Ⅰ; CITES: Appendix Ⅱ

短尾猴 *Macaca arctoides*
保护等级：中国：Ⅱ；CITES：附录Ⅱ；IUCN：VU

Protection Class: China: Ⅱ; CITES: Appendix Ⅱ; IUCN: VU

猕猴 *Macaca mulatta*
保护等级：中国：Ⅱ；CITES：附录Ⅱ

Protection Class: China: Ⅱ; CITES: Appendix Ⅱ

黑叶猴 *Presbytis francoisi*
保护等级：中国：Ⅰ；IUCN：VU

Protection Class: China: Ⅰ; IUCN: VU

食肉类哺乳动物 (Flesh-eating Mammals)

食肉目动物包括犬科、熊科、鼬科、猫科等。它们的共同特点是主要以捕食其它脊椎动物为食。与食肉习性相适应，多数种类有发达的犬齿。

Flesh-eating animals include family canidae, ursidae, mustelidae, and felidae. Their main feature is that they prey on other vertebrate. Compatible with their flesh-eating habit, most of them have developed canine teeth.

犬科动物 (Family canidae)

犬科动物四肢细长，前足五趾，后足四趾。掌小，爪直而钝，不能伸缩。耳大而直立，尾毛长而蓬松。

Family canidae animals have long and thin limbs with five toes in front feet and four toes in rear feet. The palm is small with blunt claw, which is inflexible. Their ears are large and vertical; the tails are long and fluffy.

豺 *Cuon alpinus*
保护等级：中国：Ⅱ；CITES：附录Ⅱ；IUCN：VU。

Protection Class: China: Ⅱ; CITES: Appendix Ⅱ; IUCN: VU.

狼 Wolf

狼吻部较尖，口宽阔，耳竖立；尾巴短直，始终下垂；毛色多为黄灰色或青灰色。生活在森林、草原、半荒漠、山地丘陵。主要捕食有蹄类。

Wolves have relatively narrow chins with broad mouth. Their ears are vertical; their tails are short and straight in a dropping state; the fur color is yellow grey or blue grey. They live in forest, grassland, semi-desert and mountainous regions and hills. They mainly prey on ungulate.

狼 *Canis lupus*　保护等级：CITES：附录Ⅱ
Protection Class: CITES: Appendix Ⅱ

熊科动物 (Family ursidae)

熊科动物是大型食肉动物。身体粗壮，四肢短而有力，尾短，眼小，耳小而圆。

Family ursidae are large flesh eating animals. Their bodies are big and strong, the limbs are short and powerful, the tail is short, eye small, ear tiny and round.

大熊猫 Giant Panda

大熊猫生活在中国四川、甘肃、陕西秦岭的海拔1300～3600米的高山森林中。善爬树，会游泳。主食竹类。

It lives in high mountain forest with an altitude of 1300~3600 meters in Sichuan, Gansu, Ch'in Mountain in Shanxi Province of China. They are excelled in climbing trees and can swim. They mainly eat bamboo.

棕熊 Brown bear

棕熊全身黄棕色至黑褐色，颈部有一白色领环。生活在森林中，独居，冬眠。

Brown bear body color is yellow brown or black brown. There is a white collar ring on the neck. They live in the forest alone and has the habit of hibernating.

大熊猫 *Ailuropoda melanoleuea*
保护等级：中国：Ⅰ；CITES：附录Ⅰ；IUCN：EN
Protection Class: China: Ⅰ; CITES: Appendix Ⅰ; IUCN: EN

棕熊 *Ursus arctos*
保护等级：中国：Ⅱ；CITES：附录Ⅰ
Protection Class: China: Ⅱ; CITES: Appendix Ⅰ

小熊猫 *Ailurus fulgens*
保护等级：中国：Ⅱ；CITES：附录Ⅱ；IUCN：EN
Protection Class: China: Ⅱ; CITES: Appendix Ⅱ; IUCN: EN

黑熊 Asiatic black bear

黑熊全身黑色，胸部有一"V"字形白斑。生活在山地森林中，独居。擅长爬树、游泳，能直立行走。嗅觉和听觉灵敏。冬天在树洞中处于半睡眠状态。

The black bear has an entire black body with a "V" shape white spot on the chest. It lives in mountain forest alone, they are good at climbing and swimming, they can also stand and walk. Their smell and hearing are very acute. In winter they live in tree caves in semi-sleeping state.

黑熊 *Ursus thibetamus*
保护等级：中国：Ⅱ；CITES：附录Ⅰ；IUCN：VU

Protection Class: China: Ⅱ ; CITES: Appendix Ⅰ ; IUCN: VU

北极熊 Polar Bear

在北极海岸上，北极熊是最大、最凶猛的食肉动物，经常捕食海豹。严冬，母熊在自挖的雪洞内产仔，3、4月间才走出雪洞，开始捕食。

In the sea shore of North Pole, the polar bear is the largest and fiercest animal, it often preys on seals. In the cold winter, the female bear litters in the cave made by itself. The polar bear doesn't walk out of the snow cave until March or April and began to prey on other animals.

北极熊 *Ursus maritimus*
保护等级：CITES：附录Ⅱ

Protection Class: CITES: Appendix Ⅱ

鼬科动物 (Family mustelidae)

鼬科动物四肢短，爪不能收缩。耳短而圆。大都有发达的肛腺。

Family mustelidae has short limbs with inflexible claws, their ears are short and round, most of them have fully developed anus glands.

水獭 Otter

水獭体毛短而密，呈棕黑色或咖啡色，趾间有蹼。生活在林木茂盛的河、溪、湖泊及岸边，穴居。主要捕食鱼类。

Their body fur is short and dense in brown black of coffee, there are webs between their toes. They live in rivers, steams and lake side with flourishing forests; they live in caves and prey on fish.

狼獾 Carcajou

狼獾体毛棕褐色，身体两侧有一浅棕色横带，从肩部开始至尾基汇合。生活在北温带林区。食性杂。

Carcajou body hair is brown with one shallow brown horizontal stripe on each side of the body; the stripe begins at the shoulder and meets at the tail base. They live in North Temperate Zone and are polyphagous.

水獭 *Lutra lutra*
保护等级：中国：Ⅱ；CITES：附录Ⅱ

Protection Class: China: Ⅱ; CITES: Appendix Ⅱ

狼獾 *Gulo gulo*
保护等级：中国：Ⅰ；IUCN：VU

Protection Class: China: Ⅰ; IUCN: VU

紫貂 *Martes zibellina*
保护等级：中国：Ⅰ

Protection Class: China: Ⅰ

伶鼬 *Mustela nivalis*

生物：多彩的世界

石貂 *Martes foina*
保护等级：中国：Ⅱ

Protection Class: China: Ⅱ

黄鼬 *Mustela sibirica*

艾鼬 *Mustela eversmannii*

狗獾 *Meles meles*

黄腹鼬 *Mustela kathiah*

黄喉貂 *Martes flavigula*
保护等级：中国：Ⅱ

Protection Class: China: Ⅱ

鼬獾 *Melogale moschata*

香鼬 *Mustela altaica*

灵猫科动物 (Viverrine)

体型细长，四肢短而尾长，耳小而圆；多数种类身上和尾部有条纹和斑点。

Their bodies are slim with short limbs and long tails, their ears are small and round, and most species have stripes and spot on the bodies and tails.

小灵猫 *Viverricula indica*
保护等级：中国：Ⅱ

Protection Class: China: Ⅱ

斑林狸 *Prionodon pardicolor*
保护等级：中国：Ⅱ；CITES：附录Ⅰ

Protection Class: China: Ⅱ; CITES: Appendix Ⅰ

大灵猫 *Viverra zibetha*
保护等级：中国：Ⅱ

Protection Class: China: Ⅱ

獴科动物 (Herpestidae)

全身无显著斑点或条纹，针毛粗长而蓬松；尾基粗大而尾尖渐细。

There is no obvious spot or stripe on the body, the needles shaped hair is long and fluffy, and the tail base is thick and gradually becomes thin to the end.

食蟹獴 *Herpestes urva*

猫科动物 (Catamount)

多数种类趾端具尖锐而弯曲的爪，能缩入爪鞘；犬齿长而尖锐；舌表面角质化，适于刮取骨骼上的肌肉。

Most species have sharp and curved claws which can contract into the sheath; the canine teeth are long and sharp; the tongue surface is keratinized suitable for taking the muscle on the bones.

东北虎 Siberian tiger

东北虎是猫科动物中的大型猛兽，头圆、耳短、尾长、四肢强大有力。体毛棕黄色，通体有黑色狭形横纹。

东北虎主要生活在森林和灌草丛生的地方。没有固定巢穴，活动区域大。善于游泳，但不会攀爬。视觉、听觉极为发达。主要捕食大型食草动物。

Tiger is a large beast in cat family; it has round head, short ears, long tail and powerful limbs. Its body hair is brown yellow with black narrow stripes.

Tigers mainly live in dense forest and shrubs. They do not have fixed nest, their movement range is very large. Tigers are excelled in swimming, but they cannot climb. They have very acute vision and hearing and mainly prey on large plant-eating animals.

东北虎 *Panthera tigris*
保护等级：中国：Ⅰ；CITES：附录Ⅰ；IUCN：EN

Protection Class: China: Ⅰ; CITES: Appendix Ⅰ; IUCN: EN

豹 Leopard

豹全身棕黄色，撒满古钱状花纹，又名金钱豹。生活在各种森林环境中。

The entire body is brown yellow with pattern in ancient coin, it is also called Panther and lives in various forests.

豹 *Panthera pardus*
保护等级：中国：Ⅰ；CITES：附录Ⅰ

Protection Class: China: Ⅰ; CITES: Appendix Ⅰ

雪豹 Snow Leopard

雪豹全身灰黄色，有不规则的黑环或黑斑。终年生活在雪线附近。

The entire body is grey yellow with irregular black rings or black spots. It lives near snow areas all year.

雪豹 *Panthera uncia*
保护等级：中国：Ⅰ；CITES：附录Ⅰ；IUCN：EN

Protection Class: China: Ⅰ; CITES: Appendix Ⅰ; IUCN: EN

云豹 Clouded Leopard

云豹毛色土灰色至浅黄褐色，全身布满深色大型的云纹状斑块。

The hair is dust grey or light yellow brown, the entire body is covered with large cloud shaped spots.

云豹 *Neofelis nebulosa*
保护等级：中国：Ⅱ；CITES：附录Ⅰ；IUCN：VU

Protection Class: China: Ⅱ; CITES: Appendix Ⅰ; IUCN: VU

金猫 Asiatic Golden Cat

金猫耳小而直立，眼大而圆。生活在热带、亚热带山地森林。

Ears are small and vertical; the eyes are large and round. They mainly live in tropical and semi-tropical mountainous forests.

豹猫 *Felis bengalensis*　保护等级：CITES：附录Ⅱ
Protection Class: CITES: Appendix Ⅱ

猞猁 Eurasian Lynx

猞猁两颊有下垂的长毛；耳尖有黑色耸立的簇毛；尾短，末端黑色。

There are dropping long hair under two cheeks, and there is black standing cluster hair on the tip of the year; the tail is short and the tail end is black.

猞猁 *Lynx lynx*
保护等级：中国：Ⅱ；CITES：附录Ⅱ

Protection Class: China: Ⅱ; CITES: Appendix Ⅱ

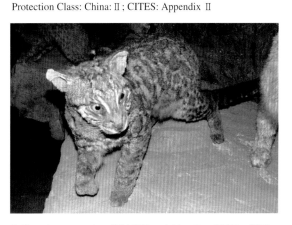

金猫 *Felis temmincki*　保护等级：中国：Ⅱ；CITES：附录Ⅰ
Protection Class: China: Ⅱ; CITES: Appendix Ⅰ

兔狲 *Felis manul*　保护等级：中国：Ⅱ；CITES：附录Ⅱ
Protection Class: China: Ⅱ; CITES: Appendix Ⅱ

狮 *Panthera leo*
保护等级：CITES：附录 II

Protection Class: CITES: Appendix II

有蹄类哺乳动物 (Ungulate Mammals)

有蹄类动物根据足趾可分奇蹄动物和偶蹄动物。奇蹄动物后肢的足趾仅第三趾发达,以此支撑身体,其余各趾退化或消失。偶蹄动物四肢的第三、第四趾发达,靠它支撑身体,其余各趾退化。

According to the number of toes the ungulate can be divided into perissodactyla and artiodactyls. Only the third toe of the rear limbs of perissodactyla is developed for supporting the body, other toes are degenerated or disappeared. The third and forth toe of artiodactyls are developed, which can support the body, other toes are degenerated.

鹿科动物 (Cervidae)

鹿科动物的角为实角,分叉,每年周期性脱换;眼下方有眶下腺;肝脏无胆囊。

The horn of Cervidae are sold horns with divarication, the horns shed yearly, there is socket gland under the eyes, there is no cholecyst in the liver.

黑麂 *Muntiacus crinifrons*
保护等级:中国:Ⅰ;CITES:附录Ⅱ;IUCN:VU

Protection Class: China: Ⅰ; CITES: Appendix Ⅱ; IUCN: VU

獐 *Hydropotes inermis*
保护等级:中国:Ⅱ

Protection Class: China: Ⅱ

黄麂 *Muntiacus reevesi*

毛冠鹿 *Elaphodus cephalophus*

驼鹿 Moose

驼鹿是世界上体型最大的鹿，肩高可达2米。其头长而大，但眼睛较小，鼻部隆厚，上唇长而肥大，喉下有一颌囊。生活在寒温带的原始针阔叶混交林中。

Moose is the largest deer in the world, the shoulder height can reach 2 m. Its head is large and long but the eyes are small, the nose is uprising and thick, the upper lip is long and fat, there is a jowl bursa under the throat. Moose mainly lives in the original mixed forest of conifer and broadleaf trees in Frigid and Temperate Zone.

驼鹿 *Alce alces*
保护等级：中国：Ⅱ。
Protection Class: China: Ⅱ.

驯鹿 Reindeer

驯鹿一般为褐色或灰色，雌雄都有角。生活在亚寒带针叶林中。

Reindeer is generally brown or grey; both male and female have horns. They live in conifer forest in Semi-frigid Zone.

驯鹿 *Rangifer tarandus*

生物：多彩的世界

马鹿 *Cervus elaphus*
保护等级：中国：II

Protection Class: China: II

马鹿 Red deer

马鹿雄性有角，一般分6叉，最多8叉。夏天呈赤褐色，冬天灰棕色。生活在高山森林或草原地区。

Male have horn, generally 6 divarications, and 8 to the most. Their body color is red brown in summer and grey brown in winter; they mainly live in high mountain forest or grassland.

梅花鹿 *Cervus nippon*
保护等级：中国：I

Protection Class: China: I

梅花鹿 Sida deer

梅花鹿雄鹿有角，角分四叉。夏毛棕褐色，遍布鲜明的白色梅花斑点，故称梅花鹿。

雄鹿的旧角每年脱落，然后长出新角。新角质地松脆，没有骨化，外面包裹着一层棕黄色的天鹅绒般的皮，皮里密布着血管，这就是著名的鹿茸。

The male deer have horns with four divarications. The hair is brown in summer dotted with bright white spots; therefore it is named spotted deer.

The horns of male deer shed every year and then the new horns grow up. The new horn is soft and crisp because it is not ossified. The outside is wrapped with a cover like velvet with dense vessels in it, this is the famous antler.

麋鹿 Milu

麋鹿是中国的特产动物,属于偶蹄目、鹿科、麋鹿属,原产于中国辽宁、华北、黄河与长江中下游。头似马、角似鹿、蹄似牛、尾似驴。体色冬天棕灰,夏天红棕;颈部至体前有一条黑褐色纵纹。雄性有角。

Milu is an animal only live in China, it belongs to milu genus, deer family artiodactyls order, it is originated in Liaoning, North China, middle and lower reaches of Yangtze River. It has a head similar to that of horse, horn similar to that of deer, hoof similar to ox and tail similar to donkey. Its body color is grey brown in winter and red brown in summer; there is a black brown vertical stripe from neck to front body. Male elk has horns.

麋鹿 *Elaphurus davidianus*
保护等级:中国Ⅰ;IUCN:CR.

Protection Class: China Ⅰ; IUCN: CR.

牛科动物 (Bovidae)

牛科动物的角为洞角，不分叉，不脱换，终身生长；肝脏有胆囊。
The horns of Bovidae are hollow horns without divarication and do not shed in the lifetime, there is cholecyst in the liver.

塔尔羊 *Hemitragus jenlahicus*
保护等级：中国 I；IUCN：VU

Protection Class: China I；IUCN: VU

北山羊 *Capra ibex*
保护等级：中国：I

Protection Class: China: I

鬣羚 *Capricornis sumatraensis*
保护等级：中国：II；CITES：附录 I；IUCN：VU

Protection Class: China: II；CITES: Appendix I；IUCN: VU

鸟类 (Birds)

鸟类具有飞翔能力，是脊椎动物中活动范围较大的一个类群，广泛分布于极地、海洋、森林、沙漠等生境。世界上已知的鸟类约有 9000 余种，中国有 1253 种 948 亚种。

Birds have the ability to fly, they are the genus having a large territory in the vertebrate, and they widely spread in polar area, ocean, forest and desert. There are totally more than 9000 kinds of birds in the world, 1253 species and 948 subspecies in China.

物种多样性展厅
The Hall of Biodiversity

鸟类的主要特征 (Main Features of Birds)

轻而坚实的骨骼 (Light but Strong Bones)

鸟类的骨骼轻巧而结实，骨骼中空无骨髓；有的骨骼内部有骨丝支架，使其更坚固；骨中的气室能够贮存足够的空气；头骨、脊柱、骨盘和肢骨的骨骼愈合；肢骨和带骨形态有较大变化等均适应飞翔。

The bones are light and strong without marrow; and some of them have bone silk bracket, which makes them stronger. The pneumatization in the bone is capable of storing enough air. The symphysis has adapted well to the flying lives, such as the skull, spine, pelvis and limb bones. In addition, the structural changings of limb and girdle bone make them fly further and live better.

1. 头骨——鸟类的头骨所有骨片完全融合连接，薄而轻
The skull is totally connected by pieces, and is very thin and light

2. 肋骨——鸟类的肋骨通过钩状突彼此相连，形成稳固的胸廓，保证胸肌的剧烈运动和完成呼吸
The ribs are connected by processus hamularis to form stable thoracic cage and to ensure the acute movement of pectoral muscle, and to accomplish the respiratory movement.

3. 胸骨——鸟类的胸骨薄而轻，中线处有龙骨突，以增大胸肌的附着面
The sternum is light and thin, the keel on the midline can enlarge the attachment surface of the pectoral muscle

4. 荐椎——愈合荐椎是鸟类特有结构，由部分胸椎、腰椎、荐椎和尾椎愈合
The synsacrum is the special skeleton part of birds, and it is composed by part of thoracic vertebra, lumbar, sacrum and caudal vertebra

5. 骨盘——开放式骨盘，与鸟类产硬壳卵的特性相适应
The open pelvis is suitable to their habit of laying hard shell eggs

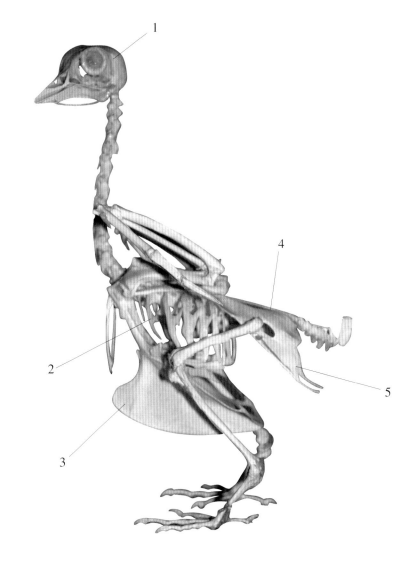

美丽的羽毛 (Beautiful Feather)

羽毛在鸟类飞行生活中发挥着极其重要的作用，比如充当飞行翼面、隔热层、动作传感器、伪装等。鸟类通常一年有两次换羽，以利于迁徙、越冬和繁殖。

Feather plays an important role in the fly of birds; it can act as wing surface, heat insulation layer, action sensor and camouflage, etc. Generally birds molt twice every year for the convenience of migration, living through the winter and reproduction.

天使之翼 (Wings of the Angel)

翅膀使得鸟类的飞行技巧富于变化，它的长度和形状取决于鸟类的食性和生活方式。
With wings, birds can fly more flexible. The length and shape of the wings depend on the feeding and living habit.

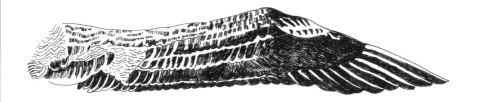

翼尖长无翼缝，适于海上滑翔，如黑脚信天翁。
The wing is long and without aperture suitable for gliding on the sea, such as black-footed albatross.

翼长宽接近，适合于急转弯、低速飞行，如雉等。
The width and length of the wings are almost the same, this kind is suitable for quick turning and low speed flying, such as pheasant, etc.

翼宽而有翼缝，适于内陆翱翔，如鹰。
Wide wings with aperture are suitable for inland hover, such as hawk.

翼末端细长，断面扁薄，翼长宽比大，适于高速飞行，多为候鸟，如燕鸥等。
The end of wings is narrow and long, and the side is flat and thin. The large ratio of wing length and width is suitable for high speed flying, and most of these birds are migratory bird such as tern.

形形色色的喙 (Various Beaks)

　　鸟喙是由上下颌骨极度前伸形成的，喙外覆坚韧的角质鞘，是鸟类区别于所有脊椎动物的结构。喙的形状和长度多种多样，每一种都与其食性和取食方式相适应的。

The beak is formed by the over-protruding of the upper and bottom jaw bones, the surface of the beak is covered by tough cutin sheath, which is the unique organ of birds in comparison with other vertebrates. The shape and length of the beaks are of various types, every kind is suitable to respective eating habit and food fetching methods.

鹈鹕（临时贮存食物） Pelican (Short-time storaging food)　　鸭（滤水） Duck (Filtering water)　　雕（撕肉） Eagle (Tearing meat)

鹬（钻泥） Snipe (Drilling mud)　　夜鹰（网飞虫） Bullbat (Netting flying insects)　　啄木鸟（凿木） Woodpecker (Chiselling wood)

各种各样的脚 (Various Feet)

　　鸟脚与其所处的生态环境是完全适应的。适于水中生活的蹼足，肉食性鸟类的脚弯曲有力，生活在寒冷地带的鸟类足趾被毛等等。

The feet are completely suitable to the environment they are living in. The palmiped is suitable for water, and bending and strong feet are for carnivorous birds, feet covered with quill-coverts are for the ones living in cold area.

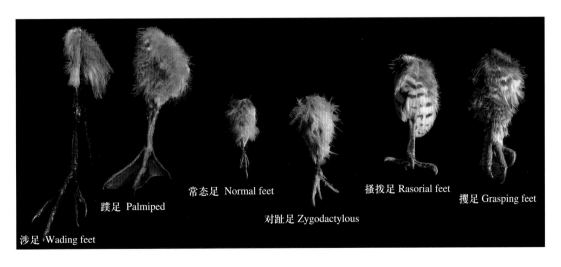

涉足 Wading feet　　蹼足 Palmiped　　常态足 Normal feet　　对趾足 Zygodactylous　　搔拨足 Rasorial feet　　攫足 Grasping feet

形形色色的鸟类 (Various Birds)

水中健将——游禽 (Swimming Birds-Master Sportsman in Water)

游禽尾脂腺发达，能分泌油脂，防止羽毛浸湿；跗跖短，趾间具蹼，适于划水；喙直尖或宽扁，有些具有嘴甲，便于捕食体滑的鱼类。

Swimming birds have fully developed oil gland which can excrete grease to prevent the feather from getting wet; they have short tarsometatarsus and there are webs between toes suitable for thrashing; the beaks are straight pointing or flat wide, some of the birds have beak shell to prey on slippery fish.

企鹅 *Spheniscus humbildti*
保护等级：CITES：附录 I

Protection Class: CITES Appendix I

黑喉潜鸟 *Gavia arctica*

大天鹅 *Cygnus cygnus*
保护等级：中国：II

Protection Class: China: II

红脚鲣鸟 *Sula sula* 栖息于热带海洋中的岛屿、海岸和海面上，善飞翔、游泳和潜水。主要以鱼类为食。中国见于西沙群岛。中国：II。

Red-footed boobys inhabits in the islands, banks and surface of tropical ocean, they are excelled in flying, swimming and diving, they mainly take fishes as their food, they live in the Xisha archipelago of China.
Protection Class: China: II

鸳鸯 *Aix galericulata* 常被作为爱情的象征而多见于文学作品，栖息于湖泊和山地森林、江河之间。中国：II。

Mandarin ducks are usually taken as the symbol of love and they ususally appear in various literatures, they inhabit in lakes, and between mountain forest and rivers.
Protection Class: China: II

中华秋沙鸭 *Mergus squamatus*
保护等级：中国：I

Protection Class: China: I

仪态万方——涉禽 (Wading Birds-With Elegant Manners)

涉禽大多在浅水滩涂处涉水生活，其胫下裸出，跗跖细长，喙和颈均较长。

Most wading birds live in shoals, their shins are bared, the tarsometatarsus is slim and long, and the beaks and neck are all relatively long.

丹顶鹤 *Grus japonensis*
栖息于湖泊、草地、海边滩涂、芦苇、沼泽等近水浅滩地带。夏季在黑龙江流域繁殖，冬季至东南沿海各省越冬。中国：Ⅰ。CITES：附录Ⅰ

Red-crowned cranes mainly inhabits in shallow water areas like lakes, grasslands, beach, reef and swamp. They mainly reproduce in Heilongjiang River area in summer, they fly to coastal areas in southeast of China.
Protection Class: China: Ⅰ, CITES Appendix Ⅰ

黑脸琵鹭 *Platalea minor*
黑脸琵鹭多在海边潮间地带及红树林和内陆水域边浅水处活动。见于中国湖南、贵州、广东、福建、台湾和海南等地。中国：Ⅱ。

Black-faced spoonbills mainly live in seaside intertidal zone, mangrove and inland shallow water area, they can be found in Hunan, Guizhou, Guangdong, Fujian, Taiwan and Hainan in China.
Protection Class: China: Ⅱ

白琵鹭 *Platalea leucorodia*
白琵鹭成群活动于河流、湖泊的浅水处，以及沿海、芦苇沼泽湿地等生境。在北方繁殖，越冬于长江下游、东南沿海及附近岛屿。中国：Ⅱ。CITES：附录Ⅱ

Platalea leucorodia They live in groups at the shallow water besides rivers and lakes, and near the sea, and in the reed swamp area. They reproduce in north, come to middle and lower reach of Yangtze River, southeast coast and nearby islands in winter.
Protection Class: China: Ⅱ, CITES Appendix Ⅱ

灰鹤 *Grus grus*
保护等级：中国：Ⅱ

Protection Class: China: Ⅱ

白枕鹤 *Grus vipio*
保护等级：CITES：附录Ⅰ

Protection Class: CITES Appendix Ⅰ

大鸨 *Otis tarda*
栖息于开阔平原、草地和半荒漠地区，以及河流、湖泊附近的干湿草地。分布于长江以北地区。中国：Ⅰ。

Great bustards inhabit in open plain, grass or semi-desert, and the dry or wet grassland near rivers and lakes. They distribute in north of Changjiang area.
Protection Class: China: Ⅰ

黑鹳 *Ciconia nigra*
常栖息于开阔的森林、湖泊、溪流、沼泽地带。中国：Ⅰ；CITES：附录Ⅱ。

Black storks usually inhabit in the open forest, lake, rivulet and swamp area.
Protection Class: China: Ⅰ, CITES Appendix Ⅱ

白鹳 *Ciconia ciconia*
栖息于开阔的平原、草地、湖泊和沼泽地带。中国见于东北和东部地区。中国：Ⅰ；CITES：附录Ⅰ。

White storks inhabit in open plain, grassland, lake and swamp area, and distribute in east and northeast area in China.
Protection Class: China: Ⅰ, CITES Appendix Ⅰ

生物：多彩的世界

黄嘴白鹭 *Egretta eulophotes*
保护等级：中国：Ⅱ

Protection Class: China: Ⅱ

大白鹭 *Egretta alba*

中白鹭 *Egretta intermedia*

金斑鸻 *Pluvialis dominica*

丘鹬 *Scolopax rusticola*

反嘴鹬 *Recurvirostra avosetta*

黑翅长脚鹬 *Himantopus himantopus*

凶猛强悍——猛禽 (Raptors-Strong and Fierce)

主要指隼形目和鸮形目的鸟类，其喙强锐而钩曲，适于撕裂肉质食物，脚强大有力，爪锐而钩曲，便于抓握，翼强大善飞翔。

Raptors include birds of falconiformes and strigiformes, their beaks are strong, sharp and curved, suitable for tearing flesh, their feet are strong and powerful, the claws are sharp and in hook shape, which is convenient for grasping, the strong wings enable their agility in flying.

金雕 *Aquila chrysaetos*
栖息于高山草原、荒漠、平原、河谷和森林地带。分布几乎遍布中国。中国：Ⅰ。
Golden eagles inhabit in plateau grassland, desert, plain area, valley and forest, they are distributed almost anywhere in China.
Protection Class: China: Ⅰ

黑翅鸢 *Elanus caeruleus*
保护等级：中国：Ⅱ
Protection Class: China: Ⅱ

雕鸮 *Bubo bubo*
保护等级：中国：Ⅱ
Protection Class: China: Ⅱ

猎隼 *Falco cherrug*
保护等级：中国：Ⅱ
Protection Class: China: Ⅱ

秃鹫 *Aegypius monachus*
保护等级：中国：Ⅱ
Protection Class: China: Ⅱ

草鸮 *Tyto capensis*
保护等级：中国：Ⅱ
Protection Class: China: Ⅱ

雪鸮 *Nyctea scandiaca*
保护等级：中国：Ⅱ
Protection Class: China: Ⅱ

快走一族——走禽 (Walking Birds--Walking So Fast)

走禽类大多胸无龙骨突，不能飞翔，但脚长而强大，善于奔走，翼不发达，可助跑。

Most walking birds do not have keel in their chests, they can not fly but have long and strong feet suitable for walking and running, the wings are not fully developed but can help them run.

鸸鹋 *Dromaius novaehollandeae*
产澳洲北部，世界第二大鸟，高1.7米，重60公斤左右。食物以各种野菜等植物为主。

The second largest bird in the world, they live in the northern part of Australia with a height of 1.7m and a weight of 60kg. They take various potherbs as food.

鸵鸟 *Struthio camelus*
产在阿拉伯及非洲北部地区，是鸟类中体型最大的，高可达2.75米，重约75-150公斤。CITES：附录Ⅰ

They live in Arab and North Africa area, and they have the largest body among all birds, they can be as tall as 2.75 m with a weight of 75-150kg.
Protection Class: CITES Appendix Ⅰ

地上行者——陆禽 (Land Birds-Walking on the Earth)

　　陆禽包括鹑鸡和鸠鸽类的鸟类，它们主要在陆地上取食种子为生，有些也吃昆虫，喙短尖，适于地面啄食，爪较钝。营地栖或树栖生活。

Land birds include partridges, pheasant and pigeons. They mainly take the seeds on the land as food, some of them also eat insects. They have short and sharp beaks suitable for pecking on land, and their claws are blunt. They live on land or trees.

白鹇 *Lophura nycthemera*
保护等级：中国：Ⅱ

Protection Class: China: Ⅱ

红腹锦鸡 *Chrysolophus pictus*
红腹锦鸡又名"金鸡"。栖于多岩山地，善于奔跑，以树芽、杂草种子和甲虫等为食。分布在青海、甘肃、贵州一带。中国：Ⅱ。

Golden pheasants inhabit in rocky mountains and are excelled in running, they take tree buds, weed seeds and beatles as food. They distribute in Qinghai, Gansu and Guizhou.
Protection Class: China: Ⅱ

褐马鸡 *Crossoptilon mantchuricum*
褐马鸡是中国特产鸟类。主要栖息于2500m以下的低山丘陵地带，除繁殖期外，常成群活动。以植物性食物为主。仅分布在河北小五台山、山西西北部和吕梁山地区。中国：Ⅰ；CITES：附录Ⅰ。

Brown-eared pheasants only exist in China. They mainly inhabit in low mountains and hill areas with a altitude less than 2500m, they live in groups except the reproducing period. They mainly take plants as their foods. They only live in Hebei Small Wutai Mountain, northwestern part of Shanxi Province and Lvliang Mountain area.
Protection Class: China: Ⅰ，CITES Appendix Ⅰ

绿孔雀 *Pavo muticus*

绿孔雀栖息于热带、亚热带常绿阔叶林和混交林中，喜在开阔地带活动。仅分布于中国云南。是珍贵的观赏鸟，数量稀少，分布区域狭窄。中国：Ⅰ；CITES：附录Ⅱ。

Green peafowls inhabit in tropical and semi-tropical braodleaf forest and mixed forest, they usually move about in open areas and are only distributed in Yunnan Province of China. Protection Class: China: Ⅰ, CITES Appendix Ⅱ

吐绶鸡 *Agriocharis ocellata*

吐绶鸡又叫"火鸡"，原产北美东部和中美洲，头颈几乎裸出，生有红色肉瘤，喉下垂有红色肉瓣，羽毛大多为金属褐色或绿色。

Turkeys are originally born in Eastern America and Central America, the neck is almost bared with red caruncle, there are red boutons under the throat. Most of their feather are metal brown or green.

毛腿沙鸡 *Syrrhaptes paradoxus*

毛腿沙鸡是典型的荒漠鸟类，跗跖底部被细鳞，跖底成垫状，适于沙漠行走。

Pallas's sandgrouses are typical desert birds with tiny squama under the tarsometatarsus, the bottom of the tarsometatarsus is in mat shape suitable for walking in the desert.

花尾榛鸡 *Bonasa bonasia*
保护等级：中国：Ⅰ

Protection Class: China: Ⅰ

岩鸽 *Columba ruestris*

攀援能手——攀禽 (Climbing Birds-Excelled in Climbing)

攀禽善于攀援，多营树栖。喙或强直或钩曲，与取食树中昆虫的习性相适应；脚短健，适于在树上进行攀援活动；翼圆或近圆，不善远距离飞翔。

Climbing birds are excelled in climbing, most of them inhabit in trees. The beaks are straight or hooked suitable for eating insects in the trees; their feet are short and strong suitable for climbing on the tree; their wings are circle or nearly circle, thus they cannot fly for a long distance.

竹啄木鸟 *Gecinulus grantia*

蚁䴕 *Jynx torquilla*

冠鱼狗 *Megaceryle lugubris*

戴胜 *Upupa epops*

绯胸鹦鹉 *Psittacula alexandri*
栖息于低山和山麓常绿阔叶林中，性情温顺，易于饲养。分布于云南和两广地区。中国：Ⅱ；CITES：附录Ⅱ。

Red-breasted parakeets inhabit in low mountain and the broadleaf forest in piedmont, their temper is mild and they are very easy to be raised. They are mainly distributed in Yunnan, Guangdong and Guangxi.
Protection Class: China: Ⅱ, CITES Appendix Ⅱ

歌喉婉转——鸣禽 (Accentors-with a Beautiful Singing Voice)

鸣禽因擅长鸣叫而得名。大多喙较短，嘴端尖，脚细小。善于营巢，多成群活动。

鸣禽类善鸣的原因是其具有发达的鸣管和鸣肌。鸣管是鸟类发声的主要器官，它是位于两个支气管进入气管的交汇处的一个腔室，鸣管壁的肌肉控制控制鸣膜在腔室内的震动。因而，鸣肌越发达，便越容易控制鸣管壁的形状和紧张程度，鸣声也就更多变婉转。

Accentors are noted for their tweets. Most of them have short beaks with sharp beak ends and tiny feet. They are good at nidation and usually they move about in groups. The reason for accentors to be excelled in singing is that they have fully developed syringes and syringeal muscle. Syringes are main phonate organs of birds, the syringes locate in a cavity which is a confluence of two bronchia and the trachea, the muscle on the syringe wall controls the shake of the tympaniform membrane in the cavity, therefore, the more strong the syringeal muscle, the easier to control the shape of syringe wall and the tension degree, and the voice will be more beautiful.

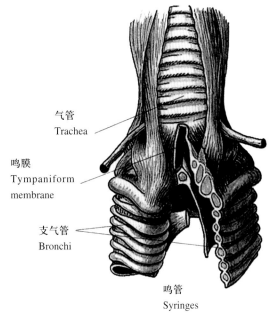

气管 Trachea
鸣膜 Tympaniform membrane
支气管 Bronchi
鸣管 Syringes

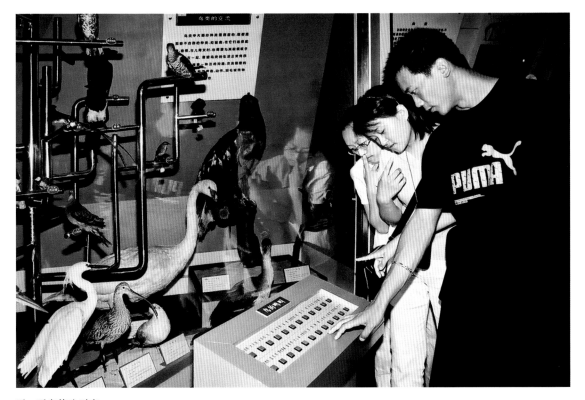

听一听鸟的鸣叫声

Listen to the voice of birds

黑枕黄鹂 *Oriolus chinensis*

冕雀 *Melanochlora sultanea*

金翅雀 *Carduelis sinica*

红嘴蓝鹊 *Urocissa erythrorhyncha*

红嘴相思鸟 *Leiothrix lutea*

极乐鸟 *Paradisaea apoda*
极乐鸟又名"天堂鸟",主要产在澳洲,栖息在热带密林中。

Birds of paradise mainly live in Australia and inhabits in tropical forest.

寿带 *Terpsiphone paradisi*

长尾阔嘴鸟 *Psarisomus dalhousiae*

北朱雀 *Carpodacus ruseus*

神奇的湿地 (Amazing Wetland)

中国湿地面积约6594万公顷，占世界湿地的10%，位居亚洲第一位，世界第四位。水鸟是湿地野生动物中最具有代表性的类群。中国有湿地水鸟12目32科271种。在亚洲57种濒危鸟类中，中国湿地内就有31种，占54%。全世界鹤类有15种，中国有记录的就有9种，占60%；全世界雁鸭类有166种，中国湿地就有50种，占30%。

The wetland in China is about 65.94 million ha, which accounts for 10% of the worlds' marsh and ranks No. 1 in Asia and No.4 in the world. Waterfowl is the typical genus in the wildlife in wetland. China has wetland waterfowls of 12 orders 32 families 271 species. Among the 57 species of endangered birds in Asia, 31 species live in China wetland, which accounts for 54% of the total number. There are totally 15 species of cranes in the world, among which 9 have recorded in China, a proportion of 60%; there are totally 166 species of ducks in the world, among which 50 species live in the wetland of China, a proportion of 30%.

湿地展厅 The Hall of Wetlands

沼泽湿地 (Swamp Wetland)

中国的沼泽湿地主要分布于东北的三江平原、大小兴安岭、若尔盖高原及海滨、湖滨、河流沿岸等，山区多木本沼泽，平原为草本沼泽。

扎龙湿地是具有代表性的沼泽湿地，已列入国际重要湿地名录。其位于温带的松嫩平原北部，面积约21万平方公里，有鸟类230多种，其中水禽120余种。主要是鹤类，包括丹顶鹤、白鹤、白枕鹤、白头鹤、灰鹤和蓑羽鹤。

The swamp wetland of China mainly distributes in Sanjiang Plain, Daxing'anling, Xiaoxing'anling, Ruo'ergai Plateau and coastal area, lake side and river banks, wood swamp is abundant in mountain areas, and herbal swamp mostly distributes in plain area.

Zhalong is a typical swamp wetland; it has been listed in the international wetland catalogue. It locates in the north of the Songnen plain in temperate zone with an area of 210 thousand square kilometers with more than 230 species of birds, among which there are 120 kinds of waterfowls, most of them is crane, including red-crowned crane, white crane, white-naped crane, hooded crane, common crane and demoiselle crane.

芦苇、沼泽湿地 Reed Swamp Wetland

海岸滩涂 (Coastal Wetland)

中国滨海湿地主要分布于沿海的11个省区和港澳台地区。海域沿岸约有1500多条大中河流入海，形成多种生态系统，浅海滩涂生态系统便是其中一种。

China coastal wetland mainly distributes in the 11 provinces along sea and Hong Kong, Macau and Taiwan. There are over 1500 large and medium sized rivers flowing into the sea, thus formed various eco-systems, and shallow eco-system is one of them.

海岸滩涂湿地 Coastal Wetland

鸟类的迁徙、环志和保护 (The Migration, Banding of Birds and Protection)

老铁山自然保护区位于辽宁省大连市旅顺口区，面积1.7万公顷，主要保护对象为黑眉蝮蛇、蛇岛特殊生态系统和候鸟。老铁山是猛禽、鸠鸽和鸣禽类迁徙的重要通道，由于其地理位置的特殊性，春、秋鸟类迁徙季节，有200多种、上百万只候鸟在此停歇。

老铁山自然保护区 Laotie Mountain Nature Reserve

Laotieshan Nature Reserve locates at Lvshunkou District of Dalian city with an area of 17 thousand ha. the main preserving animals are *Agkistrodon saxatilis*, the special eco-system of snake island and the migration birds. Laotieshan is an important pathway for the migration of raptors, pigeons and accentor. Due to its special geological location, there are more than 200 species (more than one million in number) of birds stay here in migration season in spring and autumn.

秋冬时节东亚的候鸟主要从我国北部、日、韩等地向南迁移，经阿留申群岛、日本、台湾等，往南至菲律宾群岛、婆罗洲甚至澳洲、新西兰等地；或经我国沿海往中南半岛，往南至婆罗洲、澳洲、新西兰等。春季，候鸟飞回北方繁殖。

东亚鸟类迁徙路线图
Migration Path of Birds in Eastern Asia

In autumn and winter, migrating birds fly from northern China, Japan and Korea to south via Aleutian Islands, Japan, Taiwan to the Philippines, Borneo and even Australia and New Zealand; or fly to Central South Peninsula, Borneo, Australia and New Zealand via China. In spring these birds fly back to northern part to reproduce.

两栖动物 (Amphibians)

两栖动物是脊椎动物中种类和数量最少，分布比较狭窄的一个类群。现生的两栖动物约有4000种，中国有300余种。

Amphibian is a group of animals among vertebrate with the fewest genera and species, and they also distribute in a limited number of areas. At present, there are totally about 4,000 species of amphibian, among which more than 300 species are in China.

物种多样性厅
Species Diversity Hall

有尾两栖类 (Amphibians with Tails)

终生有尾的两栖动物，包括各种鲵和蝾螈，约有350种，分布在欧亚及美洲等地。

Amphibians with tails in the whole life time include various giant salamander and axolotl, totally about 350 species distributing in Europe, Asia and America, etc.

滇池蝾螈 *Cynops wolterstorffi*

大鲵 *Andrias davidianus*
又称"娃娃鱼",是现存最大的两栖动物。
保护等级:中国:Ⅱ

Chinese Giant Salamander is the largest amphibian alive.
Protection Class: China: Ⅱ

红瘰疣螈 *Tylototriton shanjing*
保护等级:中国:Ⅱ

Protection Class: China: Ⅱ

爪鲵 *Onychodactylus fischeri*

大凉疣螈 *Tylototriton taliangensis*
保护等级:中国:Ⅱ

Protection Class: China: Ⅱ

贵州疣螈 *Tylototriton kweichowensis*
保护等级:中国:Ⅱ

Protection Class: China: Ⅱ

无尾两栖类 (Amphibians without Tails)

蛙和蟾蜍幼体和成体区别很大，成体尾部消失，四肢发达，善于跳跃，后脚趾间有蹼。有 3500 多种。

There are great differences between tadpole from the adult of frog and toad, the tails disappear, they have strong limbs and are good at jumping, there are webs between their toes of the rear legs. There are more than 3,500 species.

棘胸蛙 *Rana spinosa*
易危两栖动物
Endangered Amphibian

哀老髭蟾 *Vibrissaphora ailaonica*

中国林蛙 *Rana chensinensis*

无足两栖类 (Amphibians without Feet)

版纳鱼螈 *Ichthyophis bannanica*

爬行动物 (Reptiles)

现生的爬行动物约有 6,000 种，中国有 310 种左右。
At present there are totally about 6,000 species of reptiles, among which 310 species are in China.

蛇 (Snakes)

闪鳞蛇 *Xenopeltis unicolor*

海南闪鳞蛇 *Xenopeltis hainanensis*

蛇岛蝮 *Gloydius shedaoensisa*

灰鼠蛇 *Ptyas korros*

金环蛇 *Bungarus fasciatus*

银环蛇 *Bungarus multicinctus*

眼镜王蛇 *Ophiophagus hannah*

黑眉锦蛇 *Elaphe taeniura*

蜥蜴 (Lizards)

蜥蜴是种类最多的爬行动物,约有 3700 多种。
Lizard is a reptile with the most species; their number is more than 3700.

大壁虎 *Gekko gecko*

蜡皮蜥 *Leiolepis reevesii*

巨蜥 *Varanus salvator*

脆蛇蜥 *Ophisaurus harti*

鳄 (Crocodiles)

鳄类是现存最大的爬行动物,分布在热带亚热带地区。
Crocodile is the largest reptile alive, they distribute in tropical and semi-tropical areas.

湾鳄 *Crocodilus porosus*
中国已绝灭、国外属濒危动物

Salt-water crocodile
Already extinct in China, it is severely endangered animal in foreign countries.

扬子鳄 *Alligator sinensis*
保护等级:中国: I
游弋在长江流域被誉为"活化石"的扬子鳄,是中国特有的珍稀野生动物。

Chinese Alligator
Protection Class:China: I
Chinese alligator lives in Yangtze River area, it is reputed as the "living fossil", and it is a rare wild animal only found in China.

龟 (Tortoises)

平胸龟 *Platysternon megacephalum*

缅甸陆龟 *Indotestudo elongata*

三线闭壳龟 *Cuora trifasciata*

放射陆龟 *Geochelone radiata*

安布闭壳龟 *Cuora amboinensis*

凹甲陆龟 *Manouria impressa*

珍稀淡水鱼类 (Rare Freshwater Fish)

不论是寒带还是热带，江河湖泊中，或是即将干涸的沼泽，甚至山涧小溪里，都有淡水鱼类的踪影。

No matter in frigid area or tropical area, in river, lake or swamp nearly dry, even the stream in the valley, we can find freshwater fish.

鳇 (*Huso dauricus*)

鳇生活于江河中下层，食无脊椎动物和小鱼。分布于黑龙江中游、乌苏里江、松花江、嫩江下游。

It lives in the lower reaches of rivers and takes invertebrate as its food. It distributes in the middle reaches of Heilongjiang River, Wusuli River, Songhuajiang River and the lower reaches of Nenjiang River.

肺鱼 (Lungfish)

肺鱼是一种与腔棘鱼类相似的淡水鱼，有很发达的肺部，部分种类即使没有水也能呼吸空气而生存。在水中，鳍能像脚一样支撑着身体。

Lungfish is a kind of freshwater fish similar to coelacanth, it has strong lung and some species of lungfish can survive in the air even if there is no water. In the water, its fins can support its body like feet.

美洲肺鱼 *Lepidosiren paradoxa* Fitzinger

其他淡水鱼 (Other Freshwater Fishes)

匙吻鲟 *Polyodon spathula*
保护等级：CITES Ⅱ

Protection Class: CITES Ⅱ

食人鲳 *Serrasalmus piraya*

四眼鱼 *Anableps tetrophehalmus* Linnaeus

香鱼 *Plecoglossou altivelis* Temminck & Schlegel

黑斑狗鱼 *Esox reicherti* Dybowski

细鳞鱼 *Brachymystax lenok* (Pallas)
保护等级：中国：Ⅱ

Protection Class: China: Ⅱ

辽宁棒花鱼 *Abbottina liaoningensis* Qin,sp.nov

昆虫 (Insects)

昆虫是动物界中最成功的一个大类群。目前，已知昆虫种类约有100万种，3亿2千5百万年前，在爬行动物，鸟类和哺乳动物飞上天之前，昆虫已经借助自身强有力的翅膀翱翔在蓝天上。

Insects are the most successful species in the animal kingdom. At present, there are totally 1 million kinds of insects. More than 325 million years ago, reptiles, birds and mammals still did not have the ability of flying, but insects can fly in the sky by their own strong wings.

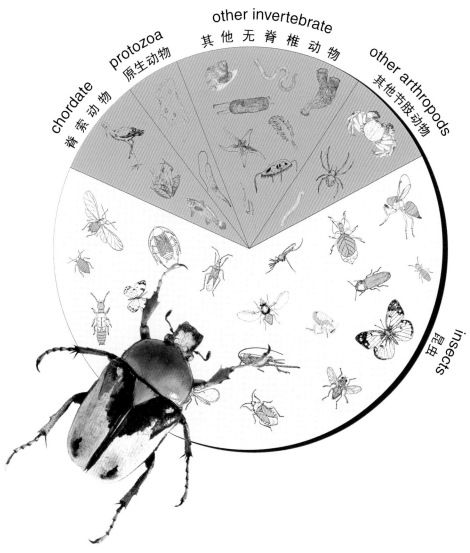

昆虫及各门类动物种群的比例结构图
Pie diagram indicating relative numbers of species of insects to the rest of the animal kingdom.

蝶与蛾 (Butterflies and Moths)

蝴蝶和蛾都属于鳞翅目昆虫，体表及膜质翅上都被有覆瓦状鳞片和毛，全球已知约17万种，蝴蝶占1/10。

Butterfly and Moth are all Lepidoptera insects, the body surface and membranous wings are covered with overlapping-tile shaped squamae and hair, there are about 170 thousand species in the world, among which 1/10 are butterfly.

蝶蛾的区别 (The difference between butterfly and moth)

蝶的触角末端膨大，棒状。
The antenna ends of butterfly are expanding and in stick shape.

蝶的鳞片呈覆瓦状排列。
The squamae of butterflies and moths are in overlapping-tile shape.

休息时蝶的两翅竖立在背上。
The two wings of buttery are upright when resting.

蓝象蛱蝶 *Perpona dexamenus*

蛾的触角形式多样，丝状、栉状。
Moth have various antennae, they can be filamentous or ctenoid.

丁香天蛾 *Psilogramma increta*

休息时蛾的翅平叠在背上。
The wings of moths are flat when resting.

粉蝶 (Pieridae)

鹤顶粉蝶 *Hebomoia glaucippe*

金顶大粉蝶 *Antes menippe*

菲色端红粉蝶 *Hebomoia leucippe*

凤蝶 (Papilionidae)

鸟翼凤蝶 *Ornithoptera priamus priamus*

鸟翼凤蝶雄蝶体长53毫米，翅展171毫米。标本采自印度尼西亚塞兰岛。

The male body length is 53mm, wingspan is 171mm; the specimen is from Ceram Island from Indonesia.

鸟翼凤蝶 *Ornithoptera priamus priamus*

多尾凤蝶 *Bhutanitis lidderdalii*

裳凤蝶 *Troides helena spilotia*

虎凤蝶 *Luehdorfia puziloi*

南美蓝带蝶 *Prepona meander*

燕凤蝶 *Lamproptera curia*

金斑喙凤蝶 *Teinopalpus aureus*

体长29毫米，翅展91毫米。中国特有珍稀蝶类，分布于海南、广东、福建等地。数量稀少。中国：I；IUCN：DD。

The body length is 29mm, wingspan is 91mm. it is a rare butterfly in China and mainly distributes in Hainan, Guangdong and Fujian, etc. The number is very limited. China: I; IUCN: DD.

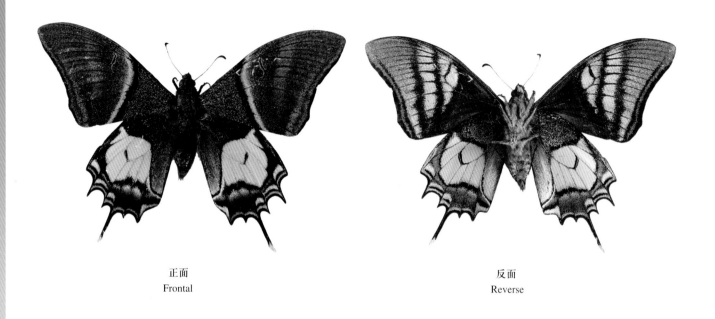

正面
Frontal

反面
Reverse

闪蝶 (Morpho)

属于鳞翅目闪蝶科（Morphidae），其拉丁名源于希腊文"Morph"一词，是美神维纳斯的名字，意味着美丽。约有80多种，全部分布在南美洲。

It belongs to Morphidae of Lepidoptera, there are totally 80 species and it mainly distributes in South America.

黎明闪蝶 *Morpho aurora*　　维纳斯闪蝶（塞浦路斯闪蝶）*Morpho cypris*　　蓝月闪蝶（大蓝闪蝶）*Morpho menelaus*

蛱蝶 (Nymphalidae)

枯叶蛱蝶 *Kallima inachus*

枯叶蛱蝶体长 34 毫米，翅展 75 毫米。是著名的拟态昆虫。休憩时，它前后翅合并，与阔叶树的枯叶极为相象，连枯叶上的主脉、次脉和霉斑都模仿得惟妙惟肖，混迹于枯叶堆中，很难发现。

Its body length is 34mm, wingspan is 75mm. It is a famous mimesis insect. Its forewings and hindwings close when resting, it seems like wizened leaves of a broadleaf tree, it even has the main nervation, sub-nervations and mildew spot on it, when it stays in the dried leaf pile, it is very hard to discover.

正面　　　　　　　　　　　　　反面
Frontal　　　　　　　　　　　　Reverse

数蛱蝶 *Diaethria aurelia*

数蛱蝶体长 15 毫米，翅展 48 毫米。翅的反面，前翅大部分是红色，端部有黑白相间的斜纹，后翅有很形象的"89"数字斑纹，蝴蝶爱好者从其身上发现了 0~9 的数字。

Its body length is 15mm, wingspan is 48mm. The inverse of the wing and the most of the forewings are red, there are tilted stripes at the wing end, there are spots on the hindwings with a pattern like the number "89", the butterfly lovers have found numbers from 0 to 9 on its body.

正面　　　　　　　　　　　　　反面
Frontal　　　　　　　　　　　　Reverse

其他蛱蝶

花色蛱蝶 *Agrias claudina*

红剑蝶 *Marpesia petreus*

缺翅安蛱蝶 *Anaea itys*

大紫蛱蝶 *Sasakia charonda*（日本国蝶）

多火炬蛱蝶 *Panacea prola*

其他蝶类 (Other Butterfly)

黑虎斑蝶 *Danaus melanippus*

箭环蝶 *Stichophthalma fruhstorferi*

尾仁眼蝶 *Corades enyo allmo*

暗神绡蝶 *Godyris duillia*

诗神袖蝶 *Heliconius melpomene*

蛾 (Moth)

乌桕大蚕蛾 *Attacus atlas*

体长42毫米，翅展186毫米，是现生最大的蛾类。分布于云南、湖南、广东、广西、福建、江西；印度，缅甸，印度尼西亚等地。

Its body length is 42mm, wingspan is 186mm. It is the largest moth alive. It distributes in Yunnan, Hunan, Guangdong, Guangxi, Fujian, Jiangxi; India, Burma and Indonesia, etc.

鬼脸天蛾 *Acherontia lachesis*

鬼脸天蛾体长47毫米，翅展121毫米。胸部背面有似脸形斑纹，面目狰狞好似"鬼脸"，故而得名。

Its body length is 47mm, wingspan is 121mm. The back of its thorax has a pattern like a face with ferocious expressions like "ghost face", so it is named like this.

其他蛾类 Other Moths

川锯翅天蛾 *Langia zenzeroides nawai*

白薯天蛾 *Herse convolvuli*

芝麻鬼脸天蛾 *Acherontia styx crathis*

樗蚕 *Samia cynthia*

绿尾大蚕蛾 *Actias selene ningpoana*

银杏大蚕蛾 *Dictyoploca japonica*

日球箩纹蛾 *Brahmophthalma japonica*

卷裳魔目夜蛾 *Eupatula macrops*

日本鹰翅天蛾 *Oxyambulyx japonica*

甲虫 (Beetles)

所有的甲虫都属于鞘翅目，包含160多个科30多万种，是动物界最大的一目。甲虫有一对厚而坚硬的前翅，即鞘翅。所有甲虫都经过从卵—幼虫—蛹—成虫的完全变态。

Beatles belongs to the coleopterous order, there are totally more than 160 families and more than 300 thousand species of it, and it is the largest order in the animal world. The beetles have a pair of thick and hard forewings, i.e. elytron. All beetles experience holometabolism like egg - larva - pupa - imago.

犀金龟类 (Dynastidae)

巨大犀金龟 *Dynastes hercules*

巨大犀金龟雄虫体长140毫米，最大可达180毫米，宽44毫米。
The male length is 140mm, the length of the largest one can reach 180mm, and the width is 44mm.

咖啡巨犀金龟 *Chalcosoma atlas*

咖啡巨犀金龟雄虫体长80～114毫米，是亚洲最大的甲虫。

The length of male is 80~114mm; it's the largest beetle in Asia.

步甲类 (Carabid Beetles)

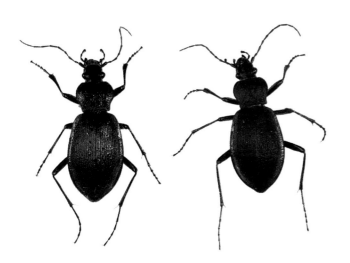
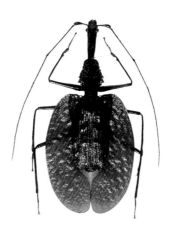

金缘蓝紫大步甲 *Tomocarabus superbulus*
体长22毫米，宽9毫米。在驼大步甲属中大多数种类是黑色的，只有朝鲜天摩山产的这个种是美丽而奇妙的蓝紫色。

Its length is 22mm and width is 9mm. Most insects of Tomocarabus is black, only this one found in Mt.Myeongjisan at Korea is beautiful blue-purple.

琴步甲 *Mormolyce phylllodes*
体长76毫米，宽35毫米。分布于东南亚热带原始林中。

The body length is 76mm and the width is 35mm, this insect mainly distributes in the wildwood in semi-tropical area in Southeast Asia.

食蜗步甲 *Damaster blaptoides*

锹甲类 (Stag Beetles)

锹甲类分类上属于鞘翅目锹甲科（Lucanidae），全球已知约有1000多种。
It belongs to the Lucanidae of Coleopterous order, more than 1000 species have been found in the world.

中国库光胫锹甲 *Odontolabis cuvera sinensis*

中国库光胫锹甲体长31.5～54毫米（不包括上颚），宽16～25.5毫米。分布于亚洲东南部山地森林。

The body length is 31.5~54mm (maxilla exclusive); the width is 16~25.5mm. They mainly distributes in the forest area in southeast mountain area in Asia.

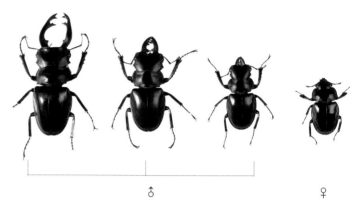

培拉玛新锹甲 *Neolucanus perarmatus*

培拉玛新锹甲体长47毫米（不包括上颚），最大79毫米，宽25毫米。已知分布于中国华南、老挝、越南北部山地原始森林中，稀有。

The maximum length is 79mm and the width is 25mm. They are found in South China, Laos, north mountain area in Viet Nam, and the insect is rare.

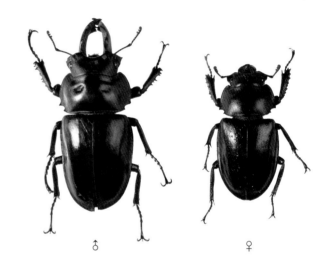

厚帕大锹甲 *Dorcus hopei hopei*

厚帕大锹甲体长32～49毫米（不包括上颚），宽13～24毫米。最大可达82毫米。中国除西藏、新疆、东北、河北、内蒙古以外，均有分布。

The maximum body length is 82mm and the width is 13~24mm. They widely distribute in China except Tibet, Sinkiang, Northeastern China, Hebei and Inner Mongolia.

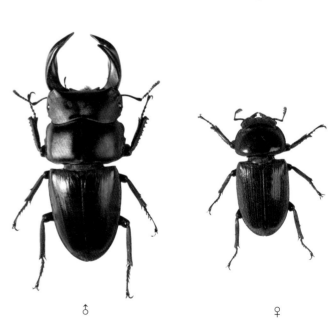

其他锹甲

巨叉锹甲 *Lucanus planeti*
生境狭窄，珍稀甲虫。

It is a rare beetle.

幸运锹甲 *Lucanida fortunei*
中国珍稀甲虫。

It is a rare beetle in China.

锹甲 *Lamrima adolphinae*

拟扁锹甲 *Dorcus consetaneus*

海尔曼深山锹甲 *Lucanus hermani*

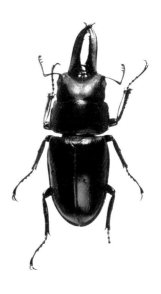

圆翅前锹甲 *Prosopocilus forficula*

吉丁虫类 (Buprestidae)

　　分类上属于鞘翅目吉丁虫科,全球已知约1万多种。成虫,白天活动,取食花、花蜜或花粉。以幼虫穿孔树木为害,常在皮下形成层部分蛀卵圆形的隧道蛀木生长。这类甲虫多分布于热带、亚热带地区的森林、果园中。

It belongs to Buprestidae of Coleopterous order; there are totally about 10 thousand species in the world.

桃紫条吉丁虫 *Chrysochroa fulgidissima*　　　吉丁虫 *Sternocera aeguisignata*　　　硕黄吉丁 *Megaloxantha bicolor*

吉丁虫 *Psiloptera castern*　　　日本吉丁虫 *Chalcophora japonica*　　　鲍氏吉丁虫 *Sternocera pulchra*
产于非洲稀有的吉丁虫。体长58毫米,宽14毫米。

It is a rare buprestid originated in Africa. Its body length is 58mm and the width is 14mm.

其他甲虫 (Other Beetles)

长臂金龟 *Euchirus macheayi*

植 物 (Plants)

　　从烈日炎炎的热带雨林，到冰封万里的极地世界，从绵延起伏的高山峡谷，到一望无际的湖泊海洋，到处都有绿色生命——植物的踪迹。

　　现存地球上的植物约 30 多万种，包括藻类植物、苔藓植物、蕨类植物、裸子植物和被子植物。

From the rain forests with burning sun to polar areas covered with ice, from the continuous mountains and valleys to the boundless lakes and seas, everywhere we can find the trace of the green lives-plants.

There are totally about more than 300 thousand kinds of plants on the earth, including alga, lichen plants, fern, gymnosperm and angiosperm.

物种多样性厅
Species Diversity Hall

东北森林厅
Northeast China Forest Hall

蕨 (Ferns)

蕨类植物是地球上最早出现的陆生植物。如今，蕨的"个子"变矮了，但它们在人们心目中的位置却越来越重，这与蕨的食用、药用价值有关，也与蕨的观赏价值有关。

Ferns are the oldest terraneous plants on the earth. At present, their "figure" has become short, but they are more and more important in our life, this is because they are edible and have officinal value, at the same time they can also be used as ornamental.

桫椤 *Alsophila spinulosa*
保护等级：中国：Ⅱ
Protection Class: China: Ⅱ

海金沙 *Lygodium japonicum*

垂穗石松 *Lycopodium cernuum*

金鸡脚 *Phymatopsis hastate*

槲蕨 *Drynaria fortunei*

翠云草 *Selaginella uncinata*

裸子植物 (Gymnosperm)

古老雄奇的家族——杉 (Old and Wonderful Family-Taxodium Family)

在距今1亿年至6千万年的地质时期，杉科树种不仅种类多，而且分布广。而今，杉科树种仅剩下16种，分布区也缩小到纬度较低的狭小区域内。

In the geological period 100 million to 60 million years ago, taxodium family has a large number of species and is widely distributed. Nowadays, only 16 species survived, and the distribution area also shrank to the narrow area in places of relatively low latitude.

水杉 *Metasequoia glyptostroboides*
保护等级：中国：Ⅰ

Protection Class: China: Ⅰ

柳杉 *Cryptomeria fortunei*

杉木 *Cunninghamia lanceolata*

濒危树种之家——红豆杉科 (Endangered Tree Family — Yew Family)

红豆杉科植物有5属23种，主要分布在北半球。中国产4属、12种，绝大多数是珍贵的保护植物。

Yew family includes 5 genera, 23 species, they mainly distribute in the north hemisphere. There are 4 genera, 12 species in China; most of them are precious plants under protection.

云南穗花杉 *Amentotaxus yunnanensis*
保护等级：中国：Ⅰ

Protection Class: China: Ⅰ

东北红豆杉 *Taxus. cuspidate* 古老的孑遗树种。
保护等级：中国：Ⅰ

Protection Class: China: Ⅰ

榧树 *Torreya grandis* 中国古老的孑遗树种。
保护等级：中国：Ⅱ

Protection Class: China: Ⅱ

被子植物 (Angiosperm)

木本经济植物之家——桦木科 (Wood Economic Plants Family-Birch Family)

桦木科植物均为落叶乔木或灌木，虽然没有著名的用材树种，但不乏造林树种、蜜源植物、食用植物和一般用材。

Plants in birch family are all defoliating arbor or shrub, there is no famous genus for practical use, but many of them can be used for forestation, as honey source plant, edible plant and general material.

华榛 *Corylus chinensis* 中国特有的优良用材和干果类珍贵树种。
保护等级：中国：Ⅲ
Protection Class: China: Ⅲ

柴桦 *Betula fruticosa* 中国东北特有植物。

千金榆 *Carpinus cordata*

香木之家——樟科 (Fragrant Family-Laurel Family)

樟科有45属、2000多种，几乎全是芳香的木本植物。樟和楠是常绿大乔木，为上等建筑和高级家具用材。

There are 45 genera, more than 2000 species in laurel family, most of them are fragrant wood plants. Camphor tree and nan are evergreen large arbor, they are superior materials for construction and furniture.

樟树 *Cinnamomum camphora*

月桂 *Laurus nobilis*

乌药 *Lindera strychnifolia*

名花良药之家——百合科 (Famous Flower and Medicine Family-Lily Family)

百合科是单子叶植物的旺族之一，有240属、4000多种，既有郁金香、风信子、百合等驰名世界的观赏花卉，也有贝母、黄精、玉竹等中国传统名药。

Lily family is one of the most thriving group in monocotyledon, it includes 240 genera, more than 4000 species, it includes both ornamental flowers like tulip, hyacinth and lily that are famous around the world and traditional Chinese medicine like fritillary, siberian solomonseal and fragrant solomonseal.

二叶舞鹤草 *Maianthemum bifolium*　　　细叶百合 *Lilium pumilum*　　　北重楼 *Paris verticillata*

鹿药 *Smilacina japonica*　　　菝葜 *Smilax china*　　　卵叶玉簪 *Hosta ensata* var. *normalis*

山花药草世家——毛茛科 (Pediment and Herd Family-Buttercup Family)

毛茛科有50属、2000多种，绝大多数是矮小的草本。

Buttercup family includes 50 genera and more than 2000 species, most of them are short herbages.

云南黄连 *Coptis teeta*
保护等级：中国：Ⅱ

Protection Class: China: Ⅱ

粉背叶人字果 *Dichocarpum hypoglaucum*
仅产云南。
保护等级：中国：Ⅱ

Protection Class: China: Ⅱ

辣蓼铁线莲 *Clematis mandshurica*

榕树之家——桑科 (Banyan Family-Mulberry Family)

桑科有40属、1000多种，其中有"独木成林"的榕树，有经济植物桑，有特殊的水果无花果，有富含乳汁的牛奶树，有结"面包"的面包树，有饮料植物啤酒花，有毒品原料大麻……

There are totally 40 genera, more than 1,000 species in mulberry family, including banyan which can "form a forest with only a single tree", economic plant mulberry, special fruit fig, milk tree with milk in it, bread tree that can produce "bread", drinking plant hop, and marijuana-the raw material of drug.

见血封喉 *Antiaris toxicaria* 树液有巨毒。
保护等级：中国：Ⅲ

Protection Class: China: Ⅲ

桑 *Morus alba*

葎草 *Humulus scandens*

奇特的有刺植物世家——小檗科 (Special Thorn Plant Family-Barberry Family)

小檗科植物有12属650种，全部是草本或灌木。

There are 12 genera, 650 species in barberry family, all of them are herbage or shrub.

八角莲 *Dysosma versipellis*
我国特有的药用观赏植物。
保护等级：中国：Ⅲ

Protection Class: China: Ⅲ

淫羊藿 *Epimedium sagittatum*

阔叶十大功劳 *Mahonia bealei*

极富盛名的花果园——蔷薇科 (Famous Flower and Fruit Garden-Rose Family)

蔷薇科在被子植物家族中极富盛名，被称为"北温带的花果园"。包括玫瑰、樱花等名花，苹果、梨、桃等佳果。

Rose family is very famous in angiosperm; it is called "the flower and fruit garden in north temperate zone". It includes famous flowers like rose and cherry blossom, and fine fruit like apple, pear and peach, etc.

香水月季 *Rosa odorata*
保护等级：中国：Ⅲ

Protection Class: China: Ⅲ

鹅绒委陵菜 *Potentilla anserine*

山楂叶悬钩子 *Rubus crataegifolius*

风箱果 *Physocarpus amurensis*

郁李 *Prunus japonica*

圆醋栗茶藨 *Ribes grossularia*

独树一帜的红叶树种——槭树 (Red-Leaf Trees with a Feature of Its Own-Maple Family)

槭树有199种，分布于亚洲、欧洲、北美洲和非洲北缘。中国是世界上槭树种类最多的国家，有151种。

There are 199 genera in Maple family, they distribute in Asia, Europe, North America and the very north of Africa. With 151 species, China has the most Maple Family species in the world.

羊角槭 *Acer yangjuechi*
保护等级：中国：II
Protection Class: China: II

色木槭 *A. mono*

元宝槭 *A. truncatum*

中国特产果树之家——无患子科 (Fruit Tree Family with Chinese Specialty -Soapberry Family)

无患子科多为乔木或灌木，有143属2000多种，我国有24属40种，包括我国特产果树龙眼和荔枝，有我国特有的单种属植物伞花木和掌叶木。

Most plants in soapberry family are arbor or shrub; there are 143 genera, more than 2000 species. There are 24 genera, 40 species in our country, among which Chinese specialties are longan and lichee, and the genus with single species such as eurycorymbus and handeliodendron can also be only found in China.

文冠果 *Xanthoceras sorbifolia*

栾树 *Koelreuteria paniculata*

无患子 *Sapindus mukorossi*

庞大的荚果之家——豆科 (Large Legumen Family-Pea Family)

豆科是被子植物的大科,有400属、10000多种。它们中有大豆、豌豆等作物,有合欢、紫荆等花木,有甘草、黄芪等药用植物,有苜蓿、草木樨等牧草。

Pea family is a large family of angiosperm, there are 400 genera, more than 10000 species, among which there are crops like soy and pea, flowers like silk tree and red bud, officinal plant like liquorice and milkvetch huangchi, and pasture like clover and yellow sweetclover.

大山黧豆 *Lathyrus davidii*

鸡眼草 *Kummerowia striata*

短梗胡枝子 *Lespedeza cyrtobotrya*

纤维植物之家——椴树科 (Fiber Plant Family-Linden Family)

椴树科有40属400多种,主产热带和亚热带。多为木本,稀为草本。茎皮富纤维,有著名的麻类作物,也有高级用材。

Linden family has 40 genera; more than 400 species mainly grow in tropical and semi-tropical area. Most of them are wood, seldom herbage. The stem is rich in fiber, there are famous jute crops and superior materials.

柄翅果 *Burretiodendron esquirolii*
保护等级:中国: Ⅱ
Protection Class: China: Ⅱ

滇桐 *Craigia yunnanensis*
保护等级:中国: Ⅱ
Protection Class: China: Ⅱ

紫椴 *Tilia amurensis*

花木草药之家——虎耳草科 (Herbal Medicine Family-Saxifrage Family)

虎耳草科有 30 属 500 多种，主产北温带地区，多为草本。

Saxifrage family includes 30 genera, more than 500 species, it mainly grows in the North Temperate Zone, and most of them are herbage.

堇叶山梅花 *Philadelphus tenuifolius*

圆锥绣球 *Hydrangea paniculata*

小花溲疏 *Deutzia parviflora*

芳香的草本之家——唇形科 (Fragrant Family-Mint Family)

唇形科有 200 属、约 3200 种，多数都是草本植物。植物体含有挥发油，具有芳香气味。

Mint family includes 200 genera, about 3,200 species; most of them are herbal plants. The plant body has naphtha in it with fragrance.

罗勒 *Ocimum basilicum*

薄荷 *Mentha haplocalyx*

连线草 *Glechoma hederacea* var. *longituba*

名贵的香料之家——龙脑香科 (Rare Flavor Family-Gurjunoiltree Family)

龙脑香科植物有17属580多种，主产于东南亚的热带地区。为名贵的香料。

Gurjunoiltree family includes 17 genera, more than 580 species and mainly grows in tropical area in Southeast Asia; most of them are rare fragrance plant.

毛叶坡垒 *Hopea mollissima*
保护等级：中国：Ⅰ

Protection Class: China: Ⅰ

望天树 *Parashorea chinensis*
保护等级：中国：Ⅰ

Protection Class: China: Ⅰ

青梅 *Vatica mangachapoi*
保护等级：中国：Ⅱ

Protection Class: China: Ⅱ

双子叶植物之冠——菊科 (The Champion of Dicotyledon Family-Composite Family)

菊科是双子叶植物种类最多的类群，有1100属、25000多种。

Composite family is the plant group with the largest number of species in dicotyledon, it includes 1,100 genera, more than 25,000 species.

栌菊木 *Nouelia insignis* 中国特有的菊科中的木本残遗种。
保护等级：中国：Ⅱ

Protection Class: China: Ⅱ

关苍术 *Atractylodes japonica*

熊耳草 *Ageratum conyzoides*

展览布局图
Exhibit layout

1F

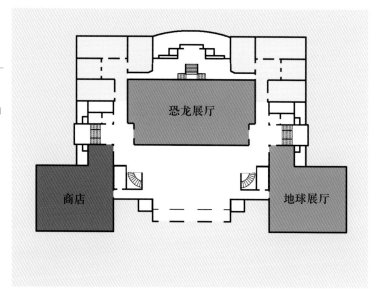

- The Hall of Dinosaur
- Shop
- The Hall of Earth

2F

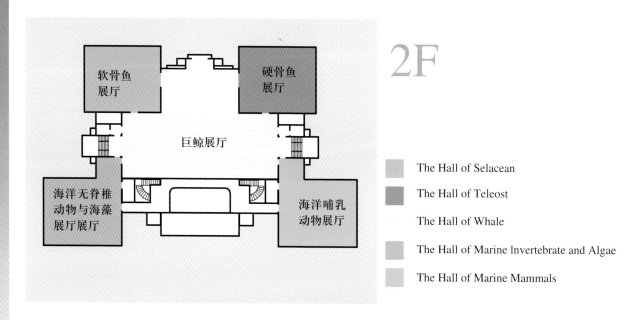

- The Hall of Selacean
- The Hall of Teleost
- The Hall of Whale
- The Hall of Marine Invertebrate and Algae
- The Hall of Marine Mammals

3F

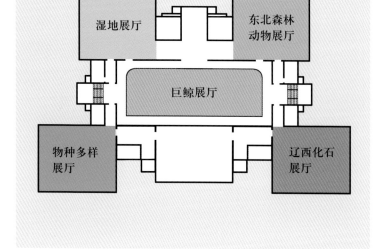

- The Hall of Wetland
- Northeast Forest Animal Hall
- The Hall of Whale
- The Hall of Western LiaoNing Fossils
- The Hall of Biodiversity

展览信息 / Exhibit Information

交通示意图
Traffic instruction

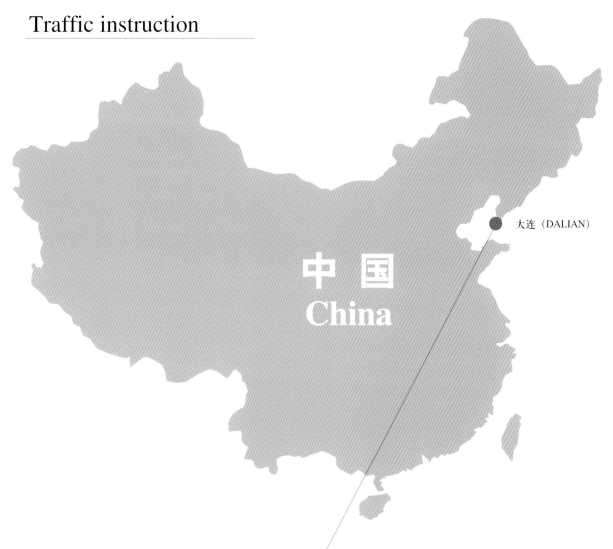

开馆时间：夏季 8：30 — 17：30
其他 9：00 — 16：30

乘车线路：23路、28路、406路、202路、
523路、801路、901路、502路、
528路、531路公交车

Open hours: Summer 8：30 — 17：30
Others   9：00 — 16：30

Bus lines: Bus No. 23, 28, 406, 202, 523, 801, 901, 502, 528, 531

后　记

　　本书是在大连市财政局的支持下，由大连自然博物馆主持完成的。国家文物局单霁翔局长为本书写了序言，对他热情支持大连自然博物馆的发展，我们深表感谢。文物出版社的苏士澍社长，对此书的出版也给予了诸多的关心、支持和协助；栾海涛先生对本书英文翻译进行审定，在此仅表谢意。同时，我们也向长期以来给予我们巨大支持的大连市科技局、全国文博界的同行、社会收藏界人士表示衷心的感谢。

　　本书除了编撰人员之外，吕义琴、张传萍、李崇和、张和春、张成富参与了照片拍摄工作；王旭日、沈才智参与翻译校对。

　　由于时间仓促，我们编撰的这本书会有许多遗漏和不足，希望大家提出宝贵意见。也希望此书的出版能够引起社会各界对大连自然博物馆的广泛关注和支持。

Postscript

This book is accomplished by Dalian Natural History Museum under the support of Dalian Financial Bureau. The director general of State Cultural Relics Bureau Shan Jixiang wrote a preface for this book, we are deeply grateful for his warm supporting the development of Dalian Natural History Museum. The Secretary of Cultural Relics Publishing House Su Shishu gave a lot of cares, supports and assistances to the publishing of this book; Mr. Luan Haitao verified the English edition of this book, herein we would like to express our thanks to them. Meanwhile we would also like to sincerely express our thanks to those giving us a great support for a long time including Dalian Technology Bureau, craft brothers of the nation's museum industry and the personalities of the social collection circle.

Besides the composers of this book, the following persons took part in the picture-taking work including Lu Yiqin, Zhang Chuanping, Li Chonghe, Zhang Hechun, Zhang Chengfu; Wang Xuri and Shen Caizhi are involved in translation verification.

Due to the short time this book we composed possibly has many missing and insufficient places, we hope readers could put forward your valuable opinions. We also hope that the publishing of this book could cause all circles of the society to widely pay more attention and support to Dalian Natural History Museum.

主要参考书目：

《惊异的大恐龙博》日本经济新闻社，2004

《中国辽西中生代鸟类》侯连海等，辽宁科学技术出版社，2002

《中国古鸟类图鉴》侯连海等，云南科技出版社，2000

《辽宁化石珍品》吴启成等，地质出版社，2002

《热河生物群》张弥曼等，上海科学技术出版社，2001

《探索书系：鸟类》辽宁教育出版社，2001

《鸟类的生活》 光复书局股份有限公司，1995

《生活自然文库——鸟类》罗杰·托里·彼得森与时代－生活丛书编辑合著.时代公司与科学出版社.1979

《中国野生哺乳动物》盛和林等编著．中国林业出版社，1999

《中国珍贵濒危动物》中华人民共和国濒危物种进出口管理办公室主编．上海科学技术出版社，1996

《目击者丛书自然博物馆——哺乳动物》史提夫 派克著．生活 读书 新知三联书店 英文汗声出版有限公司，1995